SAFETY CULTURE

Dedicated to safety champions striving to minimize preventable errors in aviation and healthcare

Safety Culture
Building and Sustaining a Cultural Change in Aviation and Healthcare

MANOJ S. PATANKAR
Saint Louis University, USA

JEFFREY P. BROWN
Healthcare Team Training, USA

EDWARD J. SABIN
Saint Louis University, USA

THOMAS G. BIGDA-PEYTON
Center for Adaptive Solutions, USA

ASHGATE

© Manoj S. Patankar, Jeffrey P. Brown, Edward J. Sabin, and Thomas G. Bigda-Peyton 2012

All rights reserved. No part of this publication may be reproduced, stored in a retrieval system or transmitted in any form or by any means, electronic, mechanical, photocopying, recording or otherwise without the prior permission of the publisher.

Manoj S. Patankar, Jeffrey P. Brown, Edward J. Sabin and Thomas G. Bigda-Peyton have asserted their right under the Copyright, Designs and Patents Act, 1988, to be identified as the authors of this work.

Published by
Ashgate Publishing Limited
Wey Court East
Union Road
Farnham
Surrey, GU9 7PT
England

Ashgate Publishing Company
Suite 420
101 Cherry Street
Burlington
VT 05401-4405
USA

www.ashgate.com

British Library Cataloguing in Publication Data
Safety culture : building and sustaining a cultural change
in aviation and healthcare.
 1. Industrial safety. 2. Safety education, Industrial--
 Evaluation. 3. Organizational change--Management.
 4. Organizational change--Evaluation. 5. Aeronautics--
 Safety measures--Evaluation. 6. Aeronautics--Human
 factors. 7. Medical care--Safety measures--Evaluation.
 8. Health facilities--Safety measures--Evaluation.
 9. Corporate culture--Case studies.
 I. Patankar, Manoj S., 1968-
 363.1'16-dc22

Library of Congress Cataloging-in-Publication Data
Safety culture : building and sustaining a cultural change in aviation and healthcare / by Manoj S. Patankar ... [et al.].
 p. cm.
 Includes bibliographical references and index.
 ISBN 9780754672371 (hardback) -- ISBN 9781409437710
(ebook) 1. Aeronautics--Safety measures. 2. Medical care--Safety measures. 3. Medical errors--Prevention. 4. Aircraft accidents--Prevention. 5. Corporate culture. I. Patankar, Manoj S., 1968-
 TL553.5.S293 2011
 363.12'46--dc23

ISBN 9780754672371 (hbk)
ISBN 9781409437710 (ebk)

2011024832

Printed and bound in Great Britain by the
MPG Books Group, UK.

Contents

List of Figures   *vii*
List of Tables   *ix*
About the Authors   *xi*
Acknowledgments   *xiii*
Foreword   *xv*
Preface   *xvii*

1	The Safety Culture Pyramid	1
2	Safety Culture Assessment	25
3	Safety Performance	57
4	Safety Climate	85
5	Safety Strategies	113
6	Safety Values	147
7	Safety Culture Transformation	179
8	Conclusion	191

Index   237

List of Figures

1.1	The Safety Culture Pyramid	2
1.2	The dominant states of safety culture along the accountability scale	6
1.3	The dominant states of safety culture along the learning scale	8
1.4	Barriers to organizational learning	9
1.5	The safety culture continuum	16
1.6	Two-dimensional look at the Safety Culture Pyramid	21
2.1	Safety culture assessment methods	27
2.2	Partial list of NTSB recommendations regarding human fatigue	32
3.1	The Safety Culture Pyramid and the Heinrich ratio	57
3.2	Runway configuration at Blue Grass Kentucky Airport	64
3.3	Runway incursion trend by quarter	66
4.1	Conceptual illustration of factor influences in the PAC model	106
6.1	The bifurcate structure of hospitals	153
6.2	The recognition primed decision-making model	156
6.3	Results of various acculturation strategies	168
8.1	Tech Ops response to safety climate survey	199
8.2	States of safety culture in Tech Ops	201
8.3	The aviation safety action program in Tech Ops	202
8.4	The ATO Heinrich triangle	207
8.5	A statement of commitment from the chief operating officer 'the safety blueprint'	208
8.6	A statement of commitment from the VP for safety	209
8.7	Safety culture initiatives commitment	209
8.8	Crisis response scale	228

List of Tables

1.1	Barriers to learning	11
1.2	The Safety Culture Pyramid and basic organizational learning tools	12
4.1	Scale composition and reliability	105
5.1	Safety strategies maturity index	114
6.1	Underlying values, unquestioned assumptions, and strategies influencing performance	167

About the Authors

Dr. Manoj Patankar is the Executive Director of the Center for Aviation Safety Research (which he established) as well as the Vice President of Academic Affairs at Saint Louis University. Manoj has authored 50 publications, including three books, and developed research partnerships with more than 20 airlines and repair stations as well as several hospitals. He has secured total funding of more than $6 million from the Federal Aviation Administration for research in the areas of maintenance human factors and safety culture, and has established leadership in connecting flight safety with patient safety. Manoj has managed five Safety Across High-Consequence Industries conferences and in 2011 he published the first edition of the *International Journal of Safety Across High-Consequence Industries*.

Jeff Brown's work focuses on supporting improvement in the management of risk and safety in domains where there is high consequence for failure. He has supported numerous patient safety improvement efforts in United States hospitals since 1999. Prior to beginning his consulting practice in 1996, Jeff served as an administrator and faculty member for collegiate aviation education programmes. Current affiliations include the Cognitive Solutions Division of Applied Research Associates and Healthcare Team Training, LLC.

Dr. Edward Sabin is a tenured faculty member in the Department of Psychology at Saint Louis University where he serves as Director of the doctoral programme in Industrial-Organizational Psychology and Director of the Center for the Application of Behavioral Sciences. He holds a secondary appointment in the Department of Aviation Science and is a member of the Center for Aviation Safety Research. Edward's research and consultation interests include organizational assessment, change and development, organizational learning, and communication processes with a special focus on safety culture in high-consequence industries. Edward is a member of the Association for Psychological Science, the Society for Industrial and Organizational Psychology, and the Association for Aviation Psychology.

Dr. Thomas Bigda-Peyton is a consultant, researcher, and educator working across high-consequence industries such as aviation, healthcare, and workplace safety. As a practitioner-researcher for 25 years, and currently as President of Action Learning Systems in Boston, he has focused on widening and accelerating the pace of improvement in individual, organizational, and large-system change initiatives. Current programmes include the We Don't Compete on Safety Consortium (a development partnership between aviation and healthcare), the

Health Transformation Learning Partnership (an initiative designed to catalyze transformation in the Ontario healthcare system), and US 2025 (a state-level programme in the United States intended to dramatically reduce the cost of care in three states while making parallel improvements in quality, access to care, and patient safety). Tom holds a doctorate in Organizational Behaviour and Intervention from the Harvard Graduate School of Education, where he worked with two pioneers in the field of organizational learning and system dynamics, Chris Argyris and Don Schon. He also holds Master's and Bachelor's degrees from Harvard.

Acknowledgments

This book was inspired by the lessons learned through the safety culture research project sponsored by the Human Factors and Engineering Group of the Federal Aviation Administration (FAA). The project started as a simple survey research and matured into a major study in safety culture transformation. We were supported by a number of managers, employees, field personnel, and researchers. Most notably, I would like to thank the following people from the FAA: Teri Bristol, Frank DeMarco, Joan Devine, Hank Krakowski, Dino Piccione, Eddie Sierra, Richard Simmons, Richard Thoma, Beverly Williams, and Steve Zaidman. I would also like to express my deep gratitude toward hundreds of men and women across the Technical Operations group who participated in the surveys, interviews, and numerous meetings and discussions throughout the past five years of this research project.

I am grateful to Saint Louis University for supporting my research endeavours by providing me the flexibility in work schedule as well as time to travel for data collection. Several members of the University community have been incredibly supportive throughout this project. Among them, I would like to specially recognize the following: President Lawrence Biondi, S.J., the faculty and staff at Parks College of Engineering, Aviation, and Technology, and my colleagues in the Office of Vice President for Frost Campus. My travel and writing commitment have stressed all of them, and they have been very supportive and accommodating.

I would also like to thank my co-authors for their continued support and commitment toward this book. Each one of them has added a unique perspective and made this work better than it would have ever been possible. The plan changed several times throughout the writing, but it kept getting better and my colleagues remained steadfast in their commitment. I would also like to thank our editor, Guy Loft, for his strong encouragement, incredible patience, and unwavering confidence.

On the personal front, I deeply appreciate the love, support, and encouragement of my dear wife, Kirsten, and our two daughters Sanjeevani and Samriddhi. I want to specially acknowledge the sacrifices made by Samriddhi, though she may not recognize them yet (she is only five years old) because I have not been able to spend as much time with her as I would have liked. Now that the book project is complete, it's time to spend some time with the girls!

Sincerely,
Manoj

I want to thank my co-authors, and especially Manoj Patankar, for making this book a reality. I am grateful to the many health professionals, administrative and clinical, for their time spent discussing safety culture and sharing their efforts to improve patient safety in their organizations. Special thanks to Sarah Taylor for her deep insights into the world of healthcare – she has provided me with invaluable education and counsel on patient safety issues and needs over many years. And, many thanks to Don Moorman, MD for sharing his thoughts and approaches to leadership and action for the improvement of patient safety, especially in the surgical realm.
    Sincerely,
    Jeff

I would like to express my appreciation for the opportunity to work with the Federal Aviation Administration on safety culture projects. In particular, I want to thank Beverly Williams and Joan Devine, as well as the many air traffic controllers and technical operations personnel who were generous with their time and expertise to assist our research. I also want to recognize the support of my graduate students and colleagues in the Industrial-Organizational Psychology Program at Saint Louis University. Additionally, the insight, dedication, and cooperation of my coauthors, Manoj, Tom, and Jeff made this book a reality. Finally, I want to express my gratitude for the enduring love and encouragement given by my wife Bonnie and our children Lara, Christopher, Nicholas, Mary and Karin.
    Sincerely,
    Ed

I would like to thank the authors' team, and especially Manoj, for keeping us on track and pulling things together. For me, this book has emerged from an ongoing journey to bridge the worlds of organizational learning, reliability, and resilience. I have been helped immensely by my colleagues on the Steering Committee of the Safety Across High-Consequence Industries conference: Manoj Patankar, Jeff Brown, Jim Bouey, Lou Halamek, and Ken Milne, as well as my consulting colleagues, notably Marty Merry and Ted Ball. I have greatly valued the contributions of many client partners, especially Bonnie Adamson, CEO of North York General Hospital in Toronto and Linda Hummel, former Director of Patient Safety at Memorial Health System in Springfield, Illinois. I am especially grateful to my wife, Frances Bigda-Peyton, for her ongoing support during this and many other projects! Without her patience and support, my contribution to this book would not have been possible.
    Sincerely,
    Tom

Foreword

A simple but useful definition of culture is 'the way we do things here'. All of us function in multiple cultures – national, organizational, and professional – and each of these influences our behaviour in a variety of settings. Professions involving risk also have a safety culture that can be defined as a constellation of individual and group values, attitudes and behaviours that reflect an organization's commitment to safety and effective safety management.

By 1993, I had spent nearly three decades studying the performance and safety-related behaviours of aquanauts, astronauts, and pilots. My only contact with medicine had been as a consumer of healthcare services. Thus, I was surprised at a space conference in Germany when a Swiss anaesthesiologist invited me to visit the University of Basel Kantonnspital. Professor Dr. Hans-Gerhard Schaefer had come to the conclusion that the practice of effective Crew Resource Management could reduce error and increase safety in medicine. Accepting his invitation, I spent a year as Visiting Professor of Anesthesia, observing communication and teamwork in the operating theatre. (Fortunately for the patient population, I only administered anesthetics to a simulated patient, Wilhelm Tell.) My observations convinced me that the concepts of CRM could be effectively applied in medicine to enhance teamwork and culture: reducing error, increasing safety, and improving morale.

This is a long-needed volume that addresses central issues of safety in both aviation and medicine. I feel the greatest contribution Patankar and his colleagues make is the presentation of a conceptual model, the *Safety Pyramid*. The model contains four linked components, performance, climate, strategies, and values, and outlines specific procedures and practices to implement and maintain a safety culture. Its application will assist in the three critical tasks of safety – awareness, diagnosis, and action.

The information presented here will be valuable for those at all levels of both professions from student to senior management. Equally important, the model and message are relevant for those in other occupations where risk is a continuing operational reality.

Robert L. Helmreich, PhD, FRAeS
Professor Emeritus of Psychology
Director Emeritus, The University of Texas Human Factors Research Project
The University of Texas at Austin

Preface

Safety Culture is a multi-layered, dynamic concept. Most of the available literature, however, focuses on *safety climate*, which is an aggregate measure of employee attitudes and opinions regarding safety. In this book, we present the *Safety Culture Pyramid* as our common contextual framework, taking readers beyond the traditional safety climate metrics and presenting a variety of assessment and analytical tools for each level of the Pyramid. These tools will enable users to build a comprehensive description of the existing safety culture and classify it in terms of one of the temporal states. Techniques used to transform the safety culture toward the most desirable state are discussed. Finally, two comprehensive case examples are used to illustrate the application of the various tools and techniques in aviation and healthcare.

In Chapter 1, we present the *Safety Culture Pyramid* as our conceptual model. This model consists of four layers: safety performance, safety climate, safety strategies, and safety values. A review of literature and empirical evidence suggests that there are influential links among these four layers. Further, the observable characteristics of safety culture could be classified along two scales: the accountability scale and the learning scale. Conceptually, transformation of safety culture will result in desirable shifts across both scales.

In Chapter 2, we present a number of different tools that could be used to build a comprehensive description of the safety culture. For example, case analysis could be used to analyze safety performance (the tip of the Safety Culture Pyramid); survey questionnaires could be used to describe the safety climate; and artifact analysis, interviews and focus group discussions, and dialogue analysis can be used to describe the safety strategies as well as safety values. A quasi-experimental approach could be used to measure the effects of specific interventions on the transformation of safety culture.

In Chapters 3–6, we focus on each layer of the Safety Culture Pyramid, respectively. We use contemporary as well as foundational literature to develop each layer of the Safety Culture Pyramid Model and we use case examples from aviation and healthcare to illustrate the application of this model.

In Chapter 7, we discuss transformation of safety culture across the accountability scale and the learning scale. Specific strategies used to transform a safety culture across the accountability scale are discussed.

In Chapter 8, we offer two comprehensive case examples to illustrate how safety culture could be transformed through practical, reliable techniques. It is said that to change culture, we must change people because people define the culture; paradoxically, the culture also defines the people. We describe a three-step, multi-method process: pre-intervention assessment, intervention, and post-intervention assessment. The pre-intervention assessment needs to be multi-method – to discover the organizational values and underlying assumptions (values); the existing organizational structures, policies, procedures, and practices (strategies); employee attitudes and opinions (climate); and employee behaviours (performance). Different elements of the transformation efforts require different assessment techniques, but typically they involve alignment across values, strategies, climate, and performance. The post-intervention assessment needs to be multi-method as well as iterative, and it should be aimed at longitudinal measurement of change to illustrate the evolutionary, and therefore irreversible, nature of cultural change.

This book is aimed at aviation and healthcare professionals as well as organizational development students. We encourage senior managers to think of safety culture transformation as a long-term commitment, invest appropriate resources, specify the performance expectations and be prepared to modify them as necessary, be open to making the structural and procedural changes necessary for the culture to change (remove internal barriers to change), and provide the down-line managers and employees the tools necessary to implement changes. We also wish to encourage middle managers to seek commitment from senior management and participation from the down-line management and employees. We hope that organizational development students will find that the blending of theory and case discussions presented here reinforces their understanding of change management.

Respectfully,

Manoj S. Patankar,
Chesterfield, MO

Jeffrey P. Brown,
Peterborough, NH

Edward J. Sabin,
Sappington, MO

Thomas Bigda-Peyton,
Boston, MA

Chapter 1
The Safety Culture Pyramid

Introduction

Airlines and hospitals need two things to survive: financial strength and public trust. A robust safety culture bolsters both. Typically, airlines operate at a very low profit margin, teetering on the edge of bankruptcy. There are a handful of key sources of financial drain related to the management of risk and the financial consequences of incidents and accidents: direct and indirect costs of accidents, insurance costs, and fines and litigation costs attributable to preventable errors. Likewise, hospitals (many of which are non-profit organizations) operate with a very slight financial margin. Many of the same costs and sources of financial drain associated with the management of risk and consequences of mishap that apply to airlines apply to hospitals. Any efforts to reduce the financial drain contribute toward the overall strength of the organization and thereby extend its longevity. Public trust is fundamental to continued revenue. The flying public needs to trust in the safety of a particular airline just as much as the patients need to trust in the safety of care offered by a particular hospital. A purchased airline ticket is a 'promissory note' to the customer that the airline will get them to their destination safely. Similarly, admission into a hospital is a 'promissory note' that the patient will be cared for without being harmed by the processes of care – that no additional illness or injury will be inflicted that is unrelated to the condition for which the patient is being treated. Management's failure to ensure due diligence in maintaining a safe and reliable aviation/healthcare system could be grounds for accusation of negligence and consequent loss of public trust.

The material presented in this book is based on our original research in aviation and healthcare as well as on a review of pertinent literature authored by colleagues from across the world. While many of our research findings have been presented in technical reports, book chapters, conference papers, and refereed articles, this book offers a unique opportunity to present our collective work in a different, and hopefully, more interesting way. It also enables us to weave in examples of preventable errors in aviation and healthcare in a way that is not possible in traditional academic literature.

A review of safety culture research indicates use of methodologies at two levels: behavioural and attitudinal. Behavioural research in safety culture emphasized causal linkage between specific risk-prone behaviours and their outcomes. This is particularly true in the case of occupational or personal injury studies. Attitudinal research emphasizes the collective psychological state of an organization. Individual attitudes toward safety processes, safety equipment, value of safety, etc. are measured through surveys and aggregated to report collective norms. While much research has been documented in behavioural and attitudinal aspects of safety culture, little is known about the organizational structures and processes or safety strategies that may have contributed to the individual-level attitudes and behaviours. Similarly, little is known about how the underlying values and unquestioned assumptions in an organization are actualized and about how they influence employee attitudes and opinions toward safety. In order to study safety culture in a more comprehensive manner, multiple research methods are required. A combination of qualitative and quantitative tools, applied over a sufficiently long duration (longitudinal studies), together with some drawn from safety engineering, organizational psychology, and anthropology, will enable us to present a more complete assessment of safety culture. This view of safety culture, known as the *Safety Culture Pyramid*, is used in this book as a shared theoretical context that can be used to assess the existing state of safety culture, develop appropriate interventions, and measure the changes in safety culture.

The Safety Culture Pyramid

Since the appearance of the term *safety culture* in the report on the 1986 Chernobyl nuclear reactor disaster, its popularity 'has led to prolific use of this concept ahead of the science of describing and assessing it' (Patankar and Sabin, 2010). Acknowledging the multi-dimensional and dynamic nature of *safety culture,* Patankar and Sabin (2010) developed a pyramid-style conceptual model that describes it as a state of dynamic balance between four stacked layers (see Figure 1.1). At the tip of the pyramid is safety performance (or safety behaviours), followed by safety climate (or employee attitudes and opinions regarding safety), next are the safety strategies, and finally, safety values form the foundation. We present this model as a pyramid because it provides a unique way of describing the linkages across various theoretical constructs.

Figure 1.1   The Safety Culture Pyramid

Safety Performance

Historically, investigation of accidents has led to discovery of endemic problems in the organizational safety culture. Examples of such problems include blatant disregard for established procedures, lack of training, routine preference to speed-over-accuracy, and in some cases, rewarding of risk-taking behaviours. Most such investigations focus on the behavioural aspects or factors that are readily observable and directly attributable to the accident under investigation. It is very much a causal-contributory analysis. The 1989 Air Ontario accident in Dryden (Maurino, Reason, Johnston, and Lee, 1997) is perhaps the first case wherein the investigators were specifically instructed to go beyond the traditional causal mapping and uncover the deeper, organizational issues. Subsequently, accident reports have increased emphasis on organizational factors – this is evident in both the Challenger as well as the Columbia investigation reports (Vaughn, 1996; CAIB, 2003). Typically, some form of Root Cause Analysis is used to build a causal chain from the final event to all the actions and in some cases systemic circumstances leading up to the event. Once all the contributing factors are known, appropriate interventions can be developed to minimize the recurrence of a similar accident. From the perspective of the Safety Culture Pyramid, the contributing factors are very important because they may contain information about underlying and commonly present behavioural traits and systemic opportunities that should be managed in order to improve the safety performance of the organization.

Safety Climate

Safety climate, or employee attitudes and opinions regarding safety, has been measured primarily through a plethora of survey instruments. There are over 50 different survey instruments in use to measure employee attitudes and opinions regarding safety. Analyses of survey responses serve as snapshots of the prevailing attitudes and opinions. Since such responses are sensitive to the prevailing organizational conditions and employee perceptions of those conditions, such surveys are called *safety climate* surveys. If a series of safety climate surveys were conducted over multiple years, it would be possible to develop a safety climate trend across the specific factors measured by the instrument. However, most measurements have been episodic and therefore difficult to trend. Johnson (2007) used the Zohar Safety Climate Questionnaire to successfully validate safety climate as a social construct as well as a reliable predictor of safety outcomes (behaviours). Patankar and Sabin (2008) conducted a pre-intervention and post-intervention safety climate study, which demonstrated the influence of a specific intervention on safety climate. So, safety climate surveys could be used to describe longitudinal trends across various factors that are being studied or to determine the effects of specific interventions on the factors that are being studied. Nonetheless, safety climate questionnaires should be statistically validated: they should demonstrate sound validity and reliability.

Safety Strategies

We are using the term *safety strategy* as an umbrella term for organizational structures, policies, procedures, and practices, as well as leadership influence. In many aviation organizations, it is now a standard practice to have a Director of Safety, detached from the operational responsibilities of the organization and reporting directly to the CEO. Similarly, in healthcare organizations, we now have a Patient Safety Officer, who is typically housed under the VP for Quality and Safety. Internal policies and procedures, such as Standard Operating Procedures, help standardize the behavioural expectations across different individuals with similar roles and responsibilities. Industry-adopted protocols such as the Safety Management System or the Aviation/Patient Safety Reporting System serve as both performance management and benchmarking tools. Safety practices, on the other hand, are a measure of what really happens. When a troubleshooting procedure or a drug delivery procedure is particularly cumbersome or actually unsafe, the practitioners tend to develop a workaround. Such workarounds become norms and over a period of time these norms become the behaviours in practice. Vaughn (1996) calls such a trend, 'normalization of deviance'. While such a deviance is a gradual creep of practice away from published procedures, there are some examples where risk-taking behaviours are rewarded as long as they don't result in safety violations. For example, if mechanics can turn around an aircraft faster by not using personal protective equipment or pilots can save time by taxiing the aircraft faster than recommended, there are some inherent business benefits and therefore such behaviours may be either overlooked or rewarded. But when such behaviours result in accidents, the same individuals may be blamed. In a study linking safety leadership, safety climate, and safety behaviour, Wu, Chen, and Li (2008) demonstrate two paths that influence safety behaviour: one path from safety leadership to safety climate and then to safety behaviour; the other path directly from safety climate to safety behaviour. The latter path is a stronger influencer of safety behaviour; however, safety leadership is a stronger influencer of safety climate. The direct influence of safety leadership on safety climate may suggest the significance of leadership in establishing appropriate structures, policies, procedures, and safety performance expectations.

Safety Values

In high-consequence industries, such as aviation and healthcare, safety needs to be an enduring value, meaning it will be consistently practised such that it has sufficient edge over profit motives. However, rarely do we see *safety* in the list of organizational values. Instead, values are typically related to customers, market share, or humanistic elements like integrity, social responsibility, and respect. The values perceived or experienced by the employees are even more important. When an employee group was asked to list their organization's values, the dominant perception was that the organization simply wanted to make money.

In this case, 'to make money' was not explicitly listed as an espoused value of the organization, but it was certainly believed to be an enacted value. If the enacted values don't conflict with the espoused values, there's likely to be less confusion and dissatisfaction among employees, but generally the wider the gap between the espoused and enacted values, the greater the dissatisfaction among the employees. Ostroff, Kinicki, and Tamkins (2003) argue that values are owned and practised by individuals, not organizations; individuals imprint their values on the organization. So, if there's a significant gap between the espoused and enacted values of the organization, the personal values of the leadership of the organization need to be assessed. Of course, an organization may change its values over time, but such changes need to be made explicit. The link between organizational values (or leaders' values) and safety strategies tends to be more explicit in organizations where their original founders are still actively involved. However, DuPont Chemicals has managed to retain safety as its enduring corporate value since 1802. Further, it has helped its partner companies improve their safety culture as well. Clearly, there must be ways to maintain fundamental organizational values through leadership transitions.

Definition of Safety Culture

The *Safety Culture Pyramid* simply asserts that values, strategies, climate, and behaviour are linked. The notion of equilibrium asserts that these four safety culture elements may be linked to denote a variety of different temporal states. Therefore, Patankar and Sabin (2010) define safety culture as follows: 'Safety culture is a dynamically-balanced, adaptable state resulting from the configuration of values, leadership strategies, and attitudes that collectively impact safety performance at the individual, group, and enterprise level'.

Simply stated, safety culture is about 'why we do what we do'.

States of Safety Cultures

Organizational values, leadership strategies, employee attitudes and opinions, and employee behaviours are present in all organizations. They exist in a state of equilibrium: observable behaviours are a result of the dynamic balance that exists between enacted values, strategies in practice, and extant employee attitudes and opinions. Such equilibrium can be described in terms of a variety of different dominant states. Further, these states can be organized along at least two scales: the accountability scale and the learning scale. Four dominant states of safety culture along the accountability scale are as follows: 'Secretive culture, blame culture, reporting culture, and just culture' (Patankar and Sabin, 2010). Similarly, the dominant states along the learning scale are failure to learn, intermittent or isolated learning, continuous learning, and transformational learning. The specific state of safety culture along the accountability and learning scales could be determined at each level of the Safety Culture Pyramid. For example, based on the review of safety performance and reaction to specific safety-critical events, one could classify the state of safety culture along the accountability scale and/or the learning scale.

The Accountability Scale

Figure 1.2 illustrates the four states of safety culture along the accountability scale: secretive, blame, reporting, and just cultures. Regardless of the specific state of safety culture, it would be very helpful to understand how the four layers of the Safety Culture Pyramid have attained their state of equilibrium and have collectively manifested a secretive, blame, reporting, or a just culture. A deeper analysis of each layer of the Pyramid and the interaction among the four layers will help us understand why a particular state of safety culture exists.

In a Secretive Culture, the organization is highly reactive, operates in a crisis mode for most events, and basic resources are tied to operational metrics with extremely limited accommodation for safety issues. Therefore, when safety issues arise, resources are either cannibalized from existing operational commitments or external sources such as insurance claims or federal aid need to be accessed. Also, several latent failures are known to individuals; in some cases, the errors or failures are masked to prevent punitive action or loss of existing, yet extremely limited resources. As a result of chronic lack of attention to safety issues and regular operation in a crisis mode, the employee-management trust is extremely poor. In such a cultural state, the trust between local management and regional management or the headquarters is also poor.

In a Blame Culture, the organization is highly reactive. Similar to the Secretive Culture, several latent failures may be known, but they tend to remain unaddressed. The organization focuses on identification and punishment of specific individuals in investigating undesirable events; thus employee-management trust is low. Consequently, safety-related performance tends to be poor, i.e. lost-time injuries, equipment damage, avoidable medical errors, etc. tend to be high.

In a Reporting Culture, the employees are regarded as sources of critical information to prevent recurrence of undesirable safety events. Effective mechanisms exist for employees to report latent systemic failures as well as individual errors. These reports are investigated and systemic solutions are

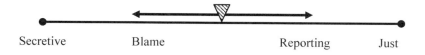

Figure 1.2 **The dominant states of safety culture along the accountability scale (Patankar and Sabin, 2010)**

implemented. Thus, employee-management trust is high and safety-related performance is strong. Examples of reporting systems include the Aviation Safety Action Program,[1] the Aviation Safety Reporting Program,[2] and the Patient Safety Reporting Program.[3]

In a Just Culture,[4] the people are encouraged, even rewarded, for providing essential safety-related information (Reason, 1997). In normal operations, emphasis is placed on the development of strong safety behaviours – actions, independent of their outcomes, are judged (Marx, 2001). Risk-taking behaviours are penalized, regardless of the actual loss/benefit, and risk-conscious safety behaviours are supported, even if they result in an undesirable event. Emphasis is placed on systemic investigations and solutions. Both management and employees are held accountable for safety improvements and therefore employee-management trust is very high.

The Learning Scale

The dominant state of the prevailing Safety Culture can be classified in terms of organizational learning literature as 'failure to learn, intermittent or isolated learning, continuous learning, and transformational learning' (Patankar and Sabin, 2008). Safety culture in aviation and healthcare can be strategically improved and transformed by engaging in organizational learning processes. Figure 1.3 illustrates the four states of safety culture along the learning scale.

Overall, organizational learning is defined as 'a process of detecting and correcting error' (Argyris, 1977). Organizational learning depends on individuals; however, individual learning alone is not sufficient to produce organizations that learn (Argyris and Schön, 1978). Crossan, Lane, and White (1999) present a useful multilevel framework for understanding the unique contributions that individuals, teams, and systems make to organizational learning. In their model, the generation of new learning begins with individual intuition and interpretation of events. Through dialogue with team members, shared understanding and joint action become possible. Lastly, institutionalization at the organizational level embeds the learning from individuals and teams into new routines, systems, rules and procedures ('strategies' in the Safety Culture Pyramid) to ensure improved performance ('safety performance or behaviours' in the Safety Culture Pyramid) in the future. In effect, such learning techniques can enable replacement of ineffective or unsafe practices with improved and safer practices, eventually changing the organizational culture.

An organization that is in a state of 'failure to learn' is characterized by recurrence of undesirable events with similar causal contributors. For example,

---

1   <http://www.faa.gov/about/initiatives/asap/>.
2   <http://www.asrs.arc.nasa.gov/>.
3   <http://www.psrs.arc.nasa.gov/flashsite/index.html>.
4   <http://www.justculture.org/default.aspx>.

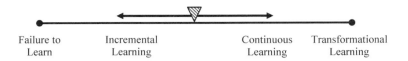

Figure 1.3   The dominant states of safety culture along the learning scale

an organization tends to have repeated heparin dosage errors due to confusing labels or repeated loss of aircraft wheels after take off due to failure to follow proper maintenance procedures. Intermittent or isolated learning typically takes place in response to specific negative experiences, and such learning is limited to preventing those specific negative experiences. For example, as a result of one case of heparin dosage error, the hospital institutes special procedures to prevent heparin dosage errors in the future, but does not address other adverse drug events with similar root cause. Similarly, in the case of aircraft maintenance, the organization conducts special training regarding wheel maintenance, but does not address known errors in other maintenance procedures with similar root cause. In contrast, an organization that is deemed to be a 'continuous learning' organization, creates systems – structures, processes, and people – that not only capture learning opportunities, but also implement solutions that address broad systemic issues. So, if a hospital is in the state of continuous learning, the heparin error would prompt that hospital to build a system that addresses all adverse drug events. Similarly, if an airline is in the state of continuous learning, the procedural error in wheel maintenance would cause the airline to build a system that reviews and improves all maintenance procedures. Finally, in the case of an organization that is in the state of 'transformational learning', the organization would already have a system in place to prevent errors and be proactive in minimizing the probability of errors across the organization. Such an organization would also be recognized among its peers as one that leads in safety innovations and shares safety information freely – an organization that does not compete on safety.

Barriers to Organizational Learning and Potential Solutions

Argyris (1994) describes how individual defensiveness and organizational defensive routines inhibit effective communication, insight, and improved action. Likewise, Mai and McAdams (1995) comment on individual and organizational impediments to learning in the form of limited perspectives (e.g. blind spots) and negative motives (e.g. fear). Senge (1990) discusses seven learning disabilities that afflict organizational settings, including the 'delusion of learning from experience'. He argues that a core learning problem is due to individuals not directly and immediately experiencing the consequences of important decisions since the results take time to become obvious. Husted and Michailova (2002) note problems of learning associated with knowledge-sharing hostility. People are not comfortable sharing knowledge and therefore may disadvantage the institution. Shaw and Perkins (1992) describe a useful model that organizes causes, symptoms, and solutions associated with barriers to organizational learning.

The Shaw and Perkins Model (Shaw and Perkins, 1992) of barriers to organizational learning is presented in Figure 1.4. The three key barriers to learning from experience are insufficiencies in reflection, action and dissemination. They argue that an individual's decisions and actions are based on a combination of cognitive, perceptual, and attitudinal factors including knowledge, expertise, opinions, strategies, values and assumptions. Events in the world are perceived and interpreted through the individual's prevailing 'belief system' which also provides the basis for action that leads to the outcomes of success or failure. Productive learning can occur when individuals or groups reflect on the consequences of their actions. Reflection that leads to insight about why things went right and turned out as expected allows successes to be sustained. Correspondingly, evaluation of differences between what was intended and what resulted allows one to learn from mistakes, errors, failures, and other unexpected or unintended outcomes. Reflection that leads to insight allows error to be detected and corrected. To be truly effective and produce a sustainable advantage, the insight needs to be disseminated, shared, understood, and integrated into the revised cultural belief system that forms the basis for improved action. For the new knowledge to be effectively applied, organizational members must be empowered to act on the revised belief system.

Shaw and Perkins (1992) suggest that symptoms of insufficient reflection include denial of problems and incomplete or incorrect analysis. This may be caused by performance pressure to produce, competency traps, and an absence of opportunities for reflection and learning. Kramer and Sabin (2003) suggest

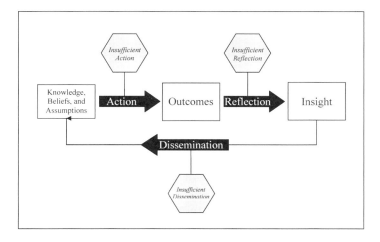

Figure 1.4 Barriers to organizational learning (adapted from Shaw and Perkins, 1992)

that managers and leaders can help overcome insufficient reflection by creating opportunities for shared reflection by implementing events such as debriefings, regular unit meetings, retreats, reflection sessions, strategic planning sessions, and learning conversations. Leaders who recognize, reward, and apply new insights will help engage organizational members in the learning process.

Ironically, Argyris (1986) points out that some of the most highly skilled individuals may in fact be the most resistant to learning and change due to what he calls skilled incompetence: 'Managers who are skilled communicators may also be good at covering up real problems' (p. 74). Argyris (1991) also argues that teaching the smartest people in the room how to learn from mistakes and failure may be the most difficult task since professionals with high levels of expertise have often been shielded from the experience of failure. Consequently, they have not experienced how to learn from failure and when things do go wrong they place blame on others rather than accepting responsibility. To correct this situation, Argyris (1991) suggests that professionals 'need to reflect critically on their own behaviour, identify the ways they often inadvertently contribute to the organization's problems, and then change how they act. In particular, they must learn how the very way they go about defining and solving problems can be a source of problems in its own right' (p. 100).

According to Shaw and Perkins (1992), symptoms of insufficient dissemination of learning include redundancy of effort and ignorance of problems and solutions. Strong inter-group boundaries may block the sharing of learning; organizational units may operate as functional information silos. Husted and Michailova (2002) discuss some of the problems and barriers associated with sharing knowledge. They conclude that reasons for hoarding information involve individual economic concerns or political power games. Recipients may reject knowledge sharing attempts due to professional pride or a desire to maintain the status quo. Cultural attitudes about whether mistakes are viewed as unavoidable or impermissible also have powerful influences on the effectiveness of knowledge dissemination. Husted and Michailova (2002) suggest several strategies for overcoming knowledge sharing hostility such as building trust and cultural expectations of knowledge sharing. Managers can also strengthen sharing and dissemination through both modelling and rewarding knowledge sharing. Shaw and Perkins (1992) suggest promoting collaboration, holding cross-functional conversations, and creating effective systems for knowledge management to promote dissemination of learning.

Knowing is not enough: action is required for organizational improvement. Shaw and Perkins (1992) discuss several reasons for a failure to act on new knowledge, including difficulties of overcoming the inertia of existing procedures, confusion among competing priorities, inadequate resources, and a sense of powerlessness to enact the required change. To encourage effective implementation of new knowledge, leaders need to be willing to challenge extant assumptions and procedures and be willing to support innovation by empowering people to act on new knowledge.

Table 1.1 summarizes the Shaw and Perkins (1992) model in the form of relationships between barriers to learning, symptoms of such barriers, causes for the barriers, and potential solutions.

According to Garvin (1993) a learning organization is 'skilled at creating, acquiring, and transferring knowledge, and at modifying its behaviour to reflect new knowledge and insights'. He argues that learning organizations are skilled at five main activities: systematic problem solving, experimentation with new approaches, learning from past experience, learning from the best practices of others, and transferring knowledge quickly and efficiently throughout the organization. In the following section, a set of learning tools are presented that can be used in aviation, healthcare, and other high-consequence industries to improve safety. These tools improve reflection, action, and dissemination of best practices. They act as countermeasures to learning disabilities by developing new learning capabilities to improve safety culture. Table 1.2 contains a summary of selected learning tools that are presented in relationship to the Safety Culture Pyramid.

Table 1.1   Barriers to learning (adapted from Shaw and Perkins, 1992)

Barrier	Symptoms	Causes	Solutions
Insufficient Reflection	• Denial of problems. • Incomplete or incorrect analysis.	• Performance pressure. • Competency traps. • Absence of learning forums.	• Stop and think. • Discuss and inquire. • Examine assumptions and experience. • Seek insight. • Think in new ways.
Insufficient Dissemination	• Problem ignorance. • Solution ignorance. • Redundancy of effort.	• Intergroup boundaries. • Myth of uniqueness. • Narrow information bandwidths.	• Collaborate. • Hold learning conversations across groups and levels. • Promote exchange and adoption of ideas across boundaries. • Build effective systems to share, store, and retrieve knowledge.
Insufficient Action	• Lack of experimentation. • Implementation failures.	• Priority stress. • Bias towards activity versus results. • Perceived powerlessness.	• Invest in experimentation and risk taking. • Empower people to act on new knowledge. • Actively apply lessons learned. • Build a learning culture.

Table 1.2   The Safety Culture Pyramid and basic organizational learning tools

Pyramid Level	Questions for Safety Leaders to Ask	Learning Tools	Purpose	Source Description
Safety Performance (behaviours)	Do we understand why mistakes and failures occur? Are we too busy to take time to reflect?	After Action Review	Improve action by understanding causes of success and failure.	Baird, Holland and Deacon (1999), Kramer and Sabin (2003)
Safety Climate (attitudes and opinions)	What aspects of our culture support or interfere with learning from experience?	Learning Capabilities Survey	Determine learning barriers and enablers.	Marsick and Watkins (2003), Goh (2003), Jerez-Gomez, et al. (2005)
Safety Strategies (mission, leadership strategies, norms, history, legends, heroes)	How do we capture and revisit key moments in our history to prepare for the future?	Learning History	Capture seminal stories that are highly instructive.	Kleiner and Roth (1997)
Safety Values (underlying values and unquestioned assumptions)	Do we practice what we preach? What topics are difficult or not discussable? Do we promote open inquiry?	Dialogue Left Hand Right Hand Column	Examine deeply held beliefs that influence critical decisions and actions.	Senge (1990), Argyris and Schön (1974)
Whole Pyramid Dynamics	Do we connect the dots? Do we see the big picture?	Systems Thinking	Understand the complex causes and structures that influence performance.	Senge (1990), Carroll (1998)

Learning Tools

Each layer of the Safety Culture Pyramid may be examined with a specific learning tool; further, all the layers may be examined in a holistic manner using 'systems thinking'. Table 1.2 illustrates the mapping of each layer with the respective questions that safety leaders must ask, a learning tool that is suited for the specific layer, the purpose of the learning tool, and sources for additional information about the tool and its application.

Many organizations continue to make the same mistakes time after time. After Action Review is a widely used learning tool that provides a powerful yet inexpensive means to examine and improve performance. It provides a systematic way to reflect on experience, determine why successes or failures occurred, and apply the lessons learned to improve future action. The intent is to fix problems not to affix blame. By determining what went wrong, similar mistakes can be avoided in the future. Likewise, by understanding what went right, effective performance can be sustained and improved. After Action Review has been used extensively by the US Army (Morrison and Meliza, 1999; Wheatley, 1994; US Army Combined Arms Center, 1993). Baird, Holland, and Deacon (1999) describe general applicability of After Action Reviews to a diverse array of organizational settings and time horizons (short-term, e.g. quick behavioural change; mid-term, e.g. policy modification; and long term, e.g. strategy transformation).

After Action Review sessions focus on five key questions: What were our objectives? What were our results – what really happened? What caused things to turn out the way they did? How do we sustain what went right? How do we improve what went wrong? The purpose of answering these questions is to uncover what the group knows following its performance that it did not know previously. Kramer and Sabin (2003) point out that for an After Action Review to be effective it must be rooted in a culture that promotes honest communication and values openness rather than defensiveness. The review should be conducted by a facilitator who frames the questions in terms of genuine dialogue that seeks to understand the causes of performance outcomes, that attempts to address underlying values, and that aims to surface unquestioned assumptions. No topics are to remain guarded from discussion. The point of the review is to capture new knowledge in lessons learned that can be quickly disseminated and acted upon. The review seeks to translate insight into improved action. An After Action Review achieves its greatest effectiveness when it becomes institutionalized as a new habit of mind and develops into a standard operating procedure for individuals, teams, and organizations.

Conducting a learning capabilities survey can determine an organization's learning barriers and enablers. Several normed surveys have been developed for this purpose. Benoit and Mackenzie (1994) describe a diagnostic process and introduce a variety of survey forms that can be used for organizational assessment. They identify 12 main enabling processes for effective learning and change that focus on the following areas: strategic direction, organizational logic, decision-making, position clarity, systematic planning, strategic human resources, innovative problem solving, enterprise-wide problem solving, performance standards, reward systems, interest compatibility, and ethical behaviour.

Goh (2003) developed a 21-item questionnaire that measures five learning capability dimensions: mission/vision clarity, leadership commitment and empowerment, experimentation and rewards, effective transfer of knowledge, and teamwork and group problem solving. The learning capability survey was used longitudinally to assess change programmes to improve learning capability in two organizations over a two to three year period. Significant improvements were documented in several dimensions of learning capability.

Jerez-Gomez, Cespedes-Lorente, and Valle-Cabrera (2005) report the development of a 16-item organizational learning capability measure. Their data were collected from a sample of chemical firms in Spain. Principal components and confirmatory factor analyses clearly supported four dimensions of organizational learning capability that included the following: managerial commitment, systems perspective, openness and experimentation, and knowledge transfer and integration.

Marsick and Watkins (2003) developed the Dimensions of the Learning Organization Questionnaire (DLOQ). This instrument collects information at the individual, team and organizational level on a total of seven dimensions: continuous learning opportunities, inquiry and dialogue, collaboration and team learning, systems to capture and share learning, collective vision, connection to the environment, and strategic leadership for learning. The scales have proven psychometrically reliable and have been used by a sizeable number of companies. Yang, Watkins, and Marsick (2004) presented a structural equation model of the DLOQ that showed significant effects of the dimensions on measures of knowledge performance and financial performance. The DLOQ is currently the most accessible and widely used measure of organizational learning. It can be used to provide insight into the learning capabilities of aviation and healthcare systems.

An organization that corrects glaring faults after a significant failure and then returns to 'business as usual' in other aspects of its operation, has learning that is intermittent at best. Intermittent learning is crisis-focused. Here results of the learning process are typically isolated to the particular event. Consequently little capacity is developed for sustained reflection on performance and it is unlikely that lessons learned will be shared with other parts of the organization. Additionally there is little interest in further performance improvement until another significant failure occurs.

By contrast, organizations that appreciate the value of continuous learning (e.g. total quality management paradigms) have defined processes in place that seek to consistently improve the quality and reliability of goods and services. Continuous learning is also described as single loop learning by Argyris (1977). Single loop learning attempts to better achieve the given goal by gradual incremental improvement.

Argyris (1977) also introduces the concept of double loop learning to describe situations where the rules themselves are questioned and individuals inquire about the goal itself and the web of assumptions related to it. Double loop learning can lead to a reframing and rethinking of the situation. More recently a newer focus has emerged that is termed triple loop learning (Wang and Ahmed, 2003). Here the focus is on learning to learn, achieving creative insight, innovation, transformation, and knowledge creation.

Learning histories capture seminal stories from the past that can be highly instructive for organizational members. The learning history makes individual experience and insight into it an organizational or systemic learning opportunity. Part of the power of the learning history comes from collective group discussion and reflection on significant events from the story. These events can be communally re-experienced and re-analyzed allowing difficult issues to be exposed for group dialogue. An important outcome is the creation of a shared understanding of the past events and of how lessons learned are relevant to the current organizational setting. Kleiner and Roth (1997) describe the process for creating the learning history document, which includes a narrative of the key events viewed from the perspective of multiple stakeholders. The document also contains commentary that seeks to uncover underlying assumptions, values, conflicts and reasoning associated with the events; and poses questions that are meant to evoke collective reflection and conversation.

Systems Thinking

Systems Thinking is an important perspective in aviation and healthcare safety because it attempts to understand the dynamics of the 'big picture' and 'connecting the dots' among multiple levels of interacting influences. The history of accident analysis in aviation demonstrates the progression towards systemic analysis. Initial attention was directed to the mechanical reliability of aircraft; however, the safety focus soon included the role and proficiency of the individual pilot. Successive generations of cockpit/crew resource management (CRM) training (e.g. Helmreich, Merritt, and Wilhelm, 1999) focused on individual competencies, group dynamic among the flight crew, interaction with other groups such as dispatchers and maintenance mechanics, integration and institutionalization of procedures into training and normal operations, and an attempt to embrace the entire system by engaging all employees to improve safety. A parallel to CRM's evolution is found in the development of a four-stage accident theory noted by Wiegmann, Zhang, von Thaden, Sharma, and Gibbons (2004). These four stages include the technical period; human error period; socio-technical period; and most recently, the organizational culture period.

The approach of systems thinking was popularized by Senge (1990) as a crucial process in organizational learning. This approach focuses on seeing the interrelationships, patterns, and underlying structures that reveal the dynamic interplay among system components. Systems thinking goes beyond typical linear analysis of cause and effect to include recognition of circular patterns of causality and feedback that play out over time and space. Senge describes a set of underlying generic structures (archetypes) that can be used to diagnose, model, and determine effective points of intervention for change. Carroll (1998) provides an example of how a systems approach can be applied in high-consequence industries.

Attempts to improve healthcare safety have increasingly turned to a systems approach to understand and reduce preventable error (e.g. Carroll and Edmondson, 2002; Larson, 2002; Johnson, Miller, and Horowitz, 2009). Vincent (2004) advocates a systems analysis to investigate clinical incidents and a systems approach to improve surgical quality and safety (e.g. Vincent, Moorthy, Sarker, Chang, and Darzi, 2004). Likewise, Waring (2007a) and argues that researchers and practitioners need to understand how contextual factors can interact in unanticipated ways in complex systems to produce accidents. However, Waring (2007b) reaches a concerning conclusion about how a physician's attribution of medical error to 'the system' can serve as a rationalization diminishing personal responsibility and allowing accommodation to a flawed system. Waring concludes that 'doctors are neither reaching out to, nor being reached by, the experts from other fields, and that the concept of "systems thinking" as promoted by the patient safety movement is not significantly penetrating the culture of frontline medical staff' (p. 45). The complexities of penetrating culture to institute systemic change can also be seen in the assessment given by Simon and Pronovost (2008), who contend that 'the publication of *To Err is Human* was the vanguard to improve patient safety. Upon nearing the report's 10-year anniversary, little appears to have changed with significant barriers encountered when attempting to track progress' (p. 2913).

Comparative Review of Safety Culture Scales

To summarize the above discussion, the state of safety culture in a given organization may be represented along two scales: the accountability scale and the learning scale. Figure 1.5 presents a conceptual parallelism across these two scales. While there is no empirical research linking the accountability scale and the learning scale, one could hypothesize that the four states across each of the scales are linked. Further, movement of along one scale may be associated with a concurrent movement along the other scale.

Aviation and Healthcare: Two High-consequence Industries Striving for a Stronger Safety Culture

Aviation and healthcare can be considered high-consequence industries because of the potential for loss of human life in the event of an error. In aviation, when the consequence of errors is catastrophic, there is national or international media attention and the entire industry suffers from negative publicity. In contrast, errors in healthcare were mostly unpublicized until the now famous Institute of Medicine report (IOM, 2001), which alarmed the public by estimating that preventable medical errors could be causing between 45,000 and 98,000 deaths per year. Comparatively, that's about five transport airliner accidents per week!

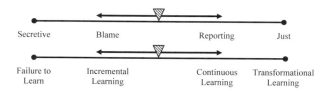

Figure 1.5 **The safety culture continuum**

In response to the 2001 IOM report, the healthcare community has engaged in numerous efforts to improve its safety culture. Since the aviation industry has been regarded as one of the High Reliability Organizations (HRO, Roberts, 1993), it was seen as an immediate ally in the quest for improved patient safety. Several successful initiatives from aviation were customized for healthcare applications. Examples of such initiatives include Crew Resource Management (CRM), Safety Climate Studies, Just Culture, Checklists, and Socio-Technical Probabilistic Risk Assessment. Saint Louis University has been hosting a series of conferences called 'Safety Across High-Consequence Industries', to provide a forum for researchers and practitioners in aviation, healthcare, nuclear power, chemical manufacturing, etc. to share their best practices, challenges, and success stories. These conferences have developed a network of safety champions who learn from each other and implement practical, effective safety solutions in their respective domains.

Patankar, Brown, Sabin, Bigda-Peyton, and Kelly (2005) reviewed safety culture literature across a variety of domains and reported three zones of safety or reliability (the terms *safety* and *reliability* are used interchangeably):

- Normal: the probability of failure is greater than one in one million operations.
- High Reliability: the probability of failure is between one in one million and one in a billion operations.
- Ultrasafe: the probability of failure is equal to, or less than, one in one billion operations.

Patankar et al. (2005) argue that the typical path from Normal to Ultrasafe is through High Reliability, which is generally achieved by extreme standardization of systems, equipment, procedures, and training. However, they also argue that the very type of standardization that allows a system to transition from Normal to High Reliability, may prevent it from transitioning further to Ultrasafe. Since Ultrasafe is becoming an essential business and societal mandate, in both aviation and healthcare, much attention has been focused on achieving that goal. Systemic improvements made in both industries, under the safety culture improvement initiatives, could help them achieve ultrasafe status.

Similarities and Differences Between Aviation and Healthcare Industries

Aviation and healthcare are similar in the overall nature of their work. Let us consider a heavy maintenance hangar in aviation. This hangar is very similar to a hospital. An airplane comes into the hangar for a 'check-up', mechanics and engineers from a variety of specialties interact with the airplane, perform their respective tasks in a predetermined manner and to pre-established standards, and finally approve the airplane for return to flight. Similarly, for inpatient services, hospitals admit a patient, physicians, nurses, and therapists provide a variety of medical, nursing, and therapy services in accordance with their respective professional standards and the needs of the patient, and finally approve the patient's discharge. In both industries, work is typically carried out in multiple shifts and over multiple days. Consequently, many of the challenges faced by aviation maintenance personnel and healthcare professionals are similar. Some examples of these challenges are as follows:

- Fatigue due to shift work or extended duty times.
- Interpersonal communication challenges due to shift-turnovers.
- Teambuilding challenges due to dynamic team membership.
- Power distance due to seniority or professional specialties.
- Speed-Accuracy Trade Offs.
- Staffing challenges.
- Professional liability.

While there are many similarities between aviation and healthcare, there are some fundamental differences as well. In the United States, healthcare does not have a single regulatory body to provide comprehensive oversight, whereas in aviation, personnel, facilities and aircraft are all certificated by the Federal Aviation Administration. Due to a lack of a comparable oversight body in healthcare, there are certain inherent challenges across the system. For example, it is difficult to standardize certain routine tasks across medical facilities. Nonetheless, organizations like the Federal Drug Administration, the Agency for Healthcare Research and Quality, as well as the Joint Commission are trying to encourage hospitals as well as medical professionals to adopt certain industry best practices. In fact, state health departments, entrusted with licensing the hospitals to conduct business, are strongly encouraging their hospitals to participate in voluntary adverse event reporting.

The fundamental nature of medical education is 'craft-based': a senior physician works with a junior physician and imparts not just the technical knowledge, but also the diagnostic heuristics. Consequently, there is a noticeable branding of physicians from one intellectual lineage over another. In aviation, the knowledge, skill, and performance expectations are more explicit, resulting in a more regimental repetition of procedures. Furthermore, the recurrent training requirements for pilots are very different from the continuing education requirements for physicians.

Medical simulation – particularly patient simulation – is relatively new. There are no explicit standards for fidelity. Comparatively, the flight simulation technology is very mature and technically advanced to the point that certain airline pilots are trained exclusively on simulators – when they fly the real airplane for the first time, it is often with a full load of passengers. Similarly, there are challenges with standardization of medical diagnostic equipment, infusion pumps, surgical room set-up, etc.

Safety Culture in Aviation

Over the last 100 years, safety culture in aviation has evolved primarily through trial and error. Since pilots are personally committed in each flight, safety of flight has a direct impact on the life of the pilots. In the early years of aviation, most of the accidents were attributed to unreliable technology; hence, research efforts were focused on improving the technology. As the hardware improved, it became more reliable; as the technical reliability improved, the range of operations expanded; as the operating range/envelope expanded, the complexity of challenges increased; and as the business of air travel became more complex, the improvements in technology alone were no longer sufficient to improve safety. Hence, in the last three decades, the emphasis has shifted toward the human element – first in terms of team communication (crew resource management and maintenance resource management) and now in terms of organizational change (safety management systems and safety culture).

Another point to bear in mind about the evolution of safety culture in aviation is the role of regulations. Aviation is arguably one of the most regulated industries. As such, safety is managed through compliance – from design, operation, and maintenance of aircraft, to initial and recurrent training of all pilots, to certification of processes and protocols used in operation and maintenance, to standardization of signage and phraseology, to certification of tools and inspection equipment, to traceability of parts all the way from raw material stock to the finished product and of course their own quality standards! So, it is fair to say that the high reliability (less than one failure in one million operations) has been attained in aviation primarily through standardization or compliance-based safety culture.

The compliance-based safety culture has reached its saturation limit: further addition of regulations is not likely to produce an appreciable increase in safety. On the contrary, addition of regulations could restrict the system's ability to improvise in the face of new threats. In the future, the safety culture needs to be based on safety as a core, enduring value that is practised as a top institutional priority, particularly when in competition with business pressures. Individual professionalism (taking responsibility to improve systemic safety through technical excellence and proactive reporting of errors and hazards) and employee-management trust (employees reporting safety issues without fear of reprisal and management taking appropriate, non-punitive systemic actions in the interest of safety) are both critical to the development of a robust safety culture.

Just Culture is an essential philosophical shift. As long as organizations and legal systems use consequence of an error, rather than the underlying behavioural pattern, as the primary criterion to decide rewards and penalties, an unjust culture will prevail. At-risk behaviours tend to be rewarded when they produce positive business outcomes and penalized when they produce negative business outcomes. In the future, emphasis must be on controlling the underlying at-risk behaviour regardless of the consequence of the behaviour. A non-punitive error reporting system, through improved quality of work and increased productivity, is a foundational mechanism to not only improve the safety, but also make a significant contribution to the financial health of the organization.

Safety Culture in Healthcare

Management of risk has always been an integral part of medical/clinical procedures. Healthcare professionals are acutely aware of risks involved in various medical procedures or clinical trials of drugs. Such awareness perhaps has also contributed to the acceptance of failures and side effects on a case-by-case basis. However, these failures and side-effects assume that there would be no preventable errors in the system. The degree to which preventable errors, such as medication errors, contribute to deaths or serious injuries was exposed in the seminal report by the Institute of Medicine (IOM), *To Err is Human*, 2001.

Since the release of the IOM report, the healthcare industry has been extensively and aggressively engaged in learning from other industries such as aviation and nuclear power. Several hospital administrators and scores of leading physicians, surgeons, and nurses have taken a proactive role in learning the state of the art regarding safety performance metrics, strategies, and best practices. Some remarkable milestones achieved by the industry in the United States include the following:

- *The National Patient Safety Foundation*[5] was formed in 1997 to 'improve the safety of the healthcare system'.
- *The National Center for Patient Safety* was established in 1999 to 'develop and nurture a culture of patient safety throughout the Veterans Health Administration'.[6]
- *The Institute for Healthcare Improvement*[7] first launched the 100,000 Lives Campaign and later the five Million Lives Campaign to bring national and international awareness as well as to build a ground swell of momentum toward reducing medical errors.
- *The Joint Commission on Accreditation of Healthcare Organizations*[8] sets specific patient safety and quality goals each year that become the goals for its accredited hospitals.
- *MORE-OB (Managing Obstetrical Risk Efficiently)* is a 'comprehensive, three-year, patient safety, professional development, and performance improvement program for caregivers and administrators in hospital obstetrics units'.[9]

---

5 <http://www.npsf.org/>.
6 <http://www.patientsafety.gov/>.
7 <http://www.ihi.org/IHI/Programs/Campaign/>.
8 <http://www.jointcommission.org/>.
9 <http://www.moreob.com>.

- *Beth Israel Deaconess Medical Center* pledged to 'eliminate all preventable harm by January 1, 2012'.[10]
- *Ascension Healthcare* recommitted itself to 'Healing Without Harm', a quest to build high-reliability healthcare.[11]

As a result of industry-wide efforts, it is now widely acknowledged that patient safety must improve and increasingly, systemic and diverse efforts are being launched across the nation.

Chapter Summary

The *Safety Culture* concept is presented in the form of a two-dimensional pyramid model: vertical and horizontal. The vertical aspect of the model presents a Safety Culture Pyramid with four layers: safety performance, safety climate, safety strategies, and safety values. Available literature suggests that these four layers are linked and that they influence each other. The horizontal aspect of the model presents safety culture in the form of temporal states that can be expressed along a *Safety Culture Continuum*, which can be delineated in terms of an accountability scale and/or a learning scale. A summative illustration of this model is presented in Figure 1.6.

Similarities and differences in aviation and healthcare are presented to lay the foundation for comparative discussion across these two industries throughout the book.

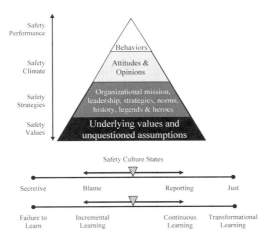

Figure 1.6 **Two-dimensional look at the Safety Culture Pyramid**

10 <http://www.bidmc.org/QualityandSafety/AspirationsforBIDMC.aspx>.
11 <http://www.ascensionhealth.org>.

References

Argyris, C. (1977). Double loop learning in organizations. *Harvard Business Review*.

Argyris, C. (1986). Skilled incompetence. *Harvard Business Review*. September–October, 74–79.

Argyris, C. (1991). Teaching smart people how to learn. *Harvard Business Review*. May-June, 99–109.

Argyris, C. (1994). Good communication that blocks learning. *Harvard Business Review*.

Argyris, C. and Schön, D. (1974). *Theory in Practice: Increasing Professional Effectiveness*. San Francisco: Jossey-Bass.

Argyris, C. and Schön, D. (1978). *Organizational Learning: A Theory of Action Perspective*. Reading: Addison Wesley.

Baird, S., Holland P. and Deacon, S. (1999). Learning from action: imbedding more learning to make a difference. *Organizational Dynamics*.

Benoit, C. and Mackenzie, K. (1994). A model of organizational learning and the diagnostic process supporting it. *The Learning Organization*, 1(3), 26–37.

CAIB (2003). Report of the Columbia accident investigation board, Volume 1. Retrieved from <http://caib.nasa.gov/news/report/volume1/default.html> on August 30, 2005.

Carroll, J. (1998). Organizational learning activities in high-hazard industries: the logics underlying self-analysis. *Journal of Management Studies*, 35(6), 669–717.

Carroll, J. and Edmondson, A. (2002). Leading organizational learning in healthcare. *Quality and Safety in Healthcare*, 11, 51–56.

Crossan, M., Lane, H. and White, R. (1999). An organizational learning framework: from intuition to institution. *Academy of Management Review*.

Garvin, D. (1993) Building a learning organization. *Harvard Business Review*.

Goh, S. (2003). Improving organizational learning capability: lessons from two case studies. *The Learning Organizaton*, 10(4), 216–227.

Helmreich, R.L., Merritt, A.C. and Wilhelm, J.A. (1999). The evolution of crew resource management training in commercial aviation. *The International Journal of Aviation Psychology*, 9, 19–32.

Husted, K. and Michailova, S. (2002). Diagnosing and fighting knowledge-sharing hostility. *Organizational Dynamics*, 31(1), 60–73.

IOM (2001). *The Institute of Medicine: Crossing the Quality Chasm: A New Health System for the 21st Century*. Washington, DC.: National Academy Press.

Jerez-Gomez, P., Cespedes-Lorente, J. and Valle-Cabrera, R. (2005). *Journal of Business Research*, 58, 715–725.

Johnson, J., Miller, S. and Horowitz, S. (2009). System-based practice: improving the safety and quality of patient care by recognizing and improving the systems in which we work. Agency for Healthcare Research and Quality <http://ahrq.hhs.gov/downloads/pub/advances2/vol2/Advances-Johnson_90.pdf>.

Johnson, S. (2007). The predictive validity of safety climate. *Journal of Safety Research*, 38, 511–521.

Kleiner, A. and Roth, G. (1997). How to make experience your company's best teacher. *Harvard Business Review*.

Kramer, T. and Sabin, E. (2003). Managing the organizational learning process to improve the bottom line. *The Psychologist-Manager Journal*, 6(1), 11–30.

Larson, E. (2002). Measuring, monitoring, and reducing medical harm from a systems perspective: a medical director's personal reflections. *Academic Medicine*, 77(10), 993–1000.

Mai, R.P. and McAdams, J.L. (1995). *Learning Partnerships: How Leading American Companies Implement Organizational Learning*. New York: McGraw-Hill.

Marsick, V. and Watkins, K. (2003). Demonstrating the value of an organization's learning culture: dimensions of the learning organization questionnaire. *Advances in Developing Human Resources*, 5(2), 132–151.

Marx, D. (2001). Patient safety and the 'just culture': a primer for healthcare executives. Retrieved from <http://www.mers-tm.net/support/Marx_Primer.pdf> on October 11, 2004.

Maurino, D., Reason, J., Johnston, N. and Lee, R. (1997). *Beyond Aviation Human Factors*. Aldershot: Ashgate.

Morrison, J. and Meliza, L. (1999). Foundations of the after action review process. Special Report 42. United States Army Research Institute for the Behavioral and Social Sciences. Alexandria, Virginia.

Ostroff, C., Kinicki, A. and Tamkins, M. (2003). Organizational culture and climate. In: W.C. Borman, D.R. Ilgen and R.J. Klimoski (eds), *Comprehensive Handbook of Psychology*, Volume 12: I/O Psychology (pp. 565–594). New York: John Wiley and Sons.

Patankar, M.S., Bigda-Peyton, T., Sabin, E., Brown, J. and Kelly, T. (2005). A comparative review of safety cultures. Report prepared for the Federal Aviation Administration. Available from <http://hf.faa.gov>.

Patankar, M.S. and Sabin, E. (2008). Safety culture transformation in technical operations of the air traffic organization: project report and recommendations. Report prepared for the Federal Aviation Administration. Available from <http://hf.faa.gov>.

Patankar, M. and Sabin, E. (2010). The safety culture perspective. In: E. Salas and D. Maurino (eds), *Human Factors in Aviation*, second edition. Chennai: Elsevier.

Reason, J. (1997). *Managing the Risk of Organizational Accidents*. Aldershot: Ashgate.

Roberts, K.H. (ed.) (1993). *New Challenges to Organizations: High Reliability Understanding Organizations*. New York: Macmillan.

Senge, P. (1990). *The Fifth Discipline: The Art and Practice of the Learning Organization*. New York: Doubleday.

Shaw, R. and Perkins, D. (1992). Teaching organizations to learn: the power of productive failures. In: D.A. Nadler, M.S. Gerstain and R.B. Shaw (eds), *Organizational Architecture: Designs for Changing Organizations*, p. 175–192. San Francisco: Josey Bass.

Simon, S. and Pronovost, P. (2008). Physician autonomy and informed decision making: finding the balance for patient safety and quality. *Journal of the American Medical Association*, 300(24), 2913–2915.

U.S. Army Combined Arms Center (1993). A leader's guide to after-action reviews (Training Circular 25–20). Fort Leavenworth, Kansas.

Vaughn, D. (1996). *The Challenger Launch Decision: Risky Technology, Culture, and Deviance at NASA*. Chicago: University of Chicago Press.

Vincent, C. (2004). Analysis of clinical incidents: a window on the system not a search for root causes. *Quality and Safety in Healthcare*, 13, 242–243.

Vincent, C., Moorthy, K., Sarker, S., Chang, A. and Darzi, A. (2004). Systems approaches to surgical quality and safety: from concept to measurement. *Annals of Surgery*, 239(4), 475–482.

Wang, C. and Ahmed, P. (2003). Organizational learning: a critical review. *The Learning Organization*, 10(1), 8–17.

Waring, J. (2007a). Getting to the 'roots' of patient safety. *International Journal for Quality in Healthcare*, 11, 1–2.

Waring, J. (2007b). Doctors' thinking about 'the system' as a threat to patient safety. *Health: An Interdisciplinary Journal for the Social Study of Health, Illness, and Medicine*, 11(1), 29–46.

Wheatley, M. (1994). Can the U.S. Army become a learning organization? *Journal for Quality and Participation*. March, 50–55.

Wiegmann, D.A., Zhang, H., von Thaden, T.L., Sharma, G. and Gibbons, A.M. (2004). Safety culture: an integrative review. *The International Journal of Aviation Psychology*, 14, 117–134.

Wu, T., Chen, C. and Li, C. (2008). A correlation among safety leadership, safety climate and safety performance. *Journal of Loss Prevention in the Process Industries*, 21, 307–318.

Yang, B., Watkins, K. and Marsick, V. (2004). The construct of the learning organization: dimensions, measurement, and validation. *Human Resource Development Quarterly*, 15(1), 31–55.

Chapter 2
Safety Culture Assessment

Introduction

In this chapter, we present an overview of various assessment techniques that may be used to build a comprehensive understanding of safety culture in a given organization. Based on the Safety Culture Pyramid Model presented in the previous chapter, we point out a repertoire of qualitative and quantitative assessment tools and techniques that will enable you to fully describe the state of safety culture in your organization.

Analysis of safety behaviours typically starts with a specific event case; hence, case analysis is the dominant technique used to analyze safety performance. In aviation, there are several tools used to analyze and report accidents. Some examples of such tools are: Human Factors Classification System (HFACS); Aviation Safety Human Reliability Analysis Method (ASHRAM); and Maintenance Error Decision Aid (MEDA). In healthcare in the United States, the most commonly recommended tools are provided by the Veterans Health Administration, the Agency for Healthcare Research and Quality, and the Institute of Safe Medication Practices. In both industries, the core methodologies incorporate the key principles of Root Cause Analysis. The causal or contributory factors, which tend to identify technical, systemic, or human factors issues, could then be used for longitudinal trending and tracking. The Heinrich Triangle or the Iceberg Model is used to illustrate how we can shift the focus from infrequently occurring fatal accidents to more frequently occurring incidents and thereby address the common underlying causes or latent systemic failures. This approach assumes that reducing the number of incidents will reduce the probability of major accidents.

As we move from safety performance to safety climate, the dominant technique is survey analysis. There is a plethora of survey instruments in use; some of them have been extensively tested for validity and reliability. We discuss some of the key scales developed through either exploratory or confirmatory factor analysis. Since safety climate is a snapshot of employee attitudes and opinions toward safety, such measurement has limited use in immediate diagnosis; however, data collected from the same organizational unit over multiple years, across different organizational units, or across different organizations in the same industry could be used to benchmark for comparative purposes, and similarities and differences across the subject populations could be studied. Pre- and post-intervention safety climate studies could be used to test the effect of specific interventions. Finally, we present some studies that have been able to demonstrate a link between safety climate and safety behaviours or performance.

Safety strategies need to be studied with a variety of tools and techniques. One of the simplest, yet quite impressive techniques, is artifact analysis. For our purpose, an artifact is physical evidence that could be used to develop a story about specific strategies (mission, values, leadership, structures, processes, policies, practices, etc.) that are used to influence the safety culture in a given organization. Other techniques include field observations like the Line Oriented Safety Audit, specific interviews or dialogues that yield stories about the history of the organization or heroes and legends in the organization, as well as interviews with 'key informants' (people who are either formal or informal leaders in the organization or those who possess deep tribal knowledge). In other words, safety strategies comprise all the things that help institutionalize the culture.

Finally, we discuss the use of artefacts, dialogue and focus group discussion to reveal the gap between espoused values and enacted values. Espoused values are the published values as they appear on corporate websites, annual reports, and general marketing materials. Enacted values are those that people working in the organization seem to experience. Generally, the tighter the alignment of values and strategies, the smaller the gap between espoused values and enacted values; the smaller the gap between espoused and enacted values, the better the safety climate; the better the safety climate, the better the safety performance.

In order to measure the effects of specific interventions on safety culture, a quasi-experimental analysis is recommended. Again, a multidisciplinary approach, borrowing from organizational psychology and anthropology, can be used. A combination of techniques for pre- and post-intervention comparisons is presented. The approach is called 'quasi' experimental because we don't have complete control over the sample population as we implement the intervention; consequently, there are factors other than the intervention itself that may have contributed to the overall change in the culture. So, the intervention and the change may not have 100 per cent cause-effect relationship.

Assessment Methods

Given the multi-faceted, multi-level nature of safety culture, both qualitative as well as quantitative methods are necessary to produce a comprehensive understanding. Figure 2.1 presents four applicable analytical techniques and their corresponding empirical results. These main methodological approaches include the following: case, survey, qualitative, and quasi-experimental analyses. Each of these approaches yields information about various layers of the Safety Culture Pyramid. Together, a multi-method approach (e.g. Di Pofi, 2002) enables triangulating of results from multiple sources (e.g. Paul, 1996) and ultimately yields a comprehensive understanding of the safety culture construct and its different states.

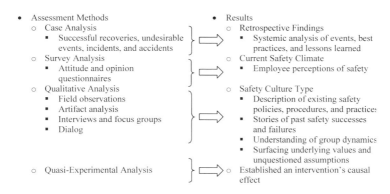

Figure 2.1   Safety culture assessment methods

Case Analysis

Behaviours of individuals and/or groups that are related to the occurrence of undesirable events, incidents or accidents attract attention from investigative and regulatory agencies. A case method is most commonly used in retrospective analysis of undesirable events as well as successful recoveries from an impending disaster (e.g. pilot Sullenberger's successful landing of US Airways flight 1549 in the Hudson river in January 2009 after a disabling bird strike).

In the early years of aviation accident investigation, the causal factors identified in retrospective case analysis were primarily technical and hence interventions were mostly focused on improving the reliability of the technology. As the aviation industry matured and its technology improved, safety attention was expanded to individual human factors and team performance issues (e.g. Crew Resource Management and Maintenance Resource Management) in addition to technical factors. Consequently, accident investigation reports reflected issues concerned with crew coordination, communication, fatigue, and adherence to policies and procedures. More recently, attention has been focused on the additional role of organizational factors in accidents, yielding a corresponding concern with organizational safety culture.

Examples of this retrospective and case-based approach in aviation include investigation techniques used by the National Transportation Safety Board and the National Aeronautics and Space Administration (e.g. NTSB, 2002; NASA, 2006). This retrospective case approach is also found in healthcare and includes Morbidity and Mortality conferences used to review medical error and undesirable patient outcomes (e.g. Gordon, 1994; Pierluissi, Fischer, Campbell, and Landefeld, 2003; and Rosenfeld, 2005) as well as investigations by the respective state-level organization overseeing medical services provided by the local hospitals (e.g. California Department of Public Health or Maine Department of Health and Human Services). Brief cases are presented to illustrate the use of Root Cause

Analysis as a technique to identify multiple causal contributors to undesirable events.

Example 2.1: Wrong-site Tooth Extraction[1]

This example is Case#156 reported on the Web-based Morbidity and Mortality database:

> A 45-year-old healthy man was scheduled to have two teeth extracted for progressive dental caries. The patient underwent the extractions, awoke from the anesthesia, and then realized that his upper left molars had been extracted instead of his right. The error was recognized and acknowledged immediately following the procedure. The patient still required extraction of the diseased teeth, which occurred a few weeks later. He developed no significant complication from either surgical procedure other than enduring two rounds of anesthesia because of the error.

In this example, the undesirable event is that upper left molars were extracted instead of upper right molars – hence, a wrong-site surgery. In a disciplinary or litigious investigation, the focus is on identifying the individual who should be held responsible for the error and punishing that individual. In the case of Root Cause Analysis (RCA), the emphasis is on building a system-wide illustration of various factors that contributed to the final error. The RCA protocol requires the investigator to validate all the facts and to keep drilling down the causal pathway – by asking a series of 'why' questions – until all rational options are exhausted, and all human errors must be preceded by a systemic condition. Why did the surgeon extract the wrong teeth? A possible response might be, 'because markings on the dental diagram were confusing'. One could continue along this pathway and discover several latent systemic issues. Dr. Smith, an expert commenter on this case, lists the following factors: 'Cognitive failure and miscommunication, multiple contiguous carious teeth (rather than one identifiable diseased tooth), partially erupted teeth mimicking third molars, teeth with gross decay that the restorative dentist wants to save, reversed radiographs, and nebulous tooth numbering systems'.

Dental diagrams,[2] such as the one provided by the Agency for Healthcare Research and Quality, may be used to standardize and structure communication between care providers, technicians, and patients to reduce the risk of wrong-site extraction. Similarly, the 'Speak-up' posters, videos, coloring books,[3] etc.

---

1   <http://www.webmm.ahrq.gov/case.aspx?caseID=156>.
2   A typical dental diagram available at <http://www.webmm.ahrq.gov/media/cases/images/case156_fig1.gif>.
3   'Speak-up' campaign details available at <http://www.jointcommission.org/speakup.aspx>.

created by the Joint Commission, could be used to raise awareness and inspire assertiveness among patients and care providers, as well as patient advocates.

The diagrams, posters, videos, slide presentations, etc. serve as artefacts memorializing and reinforcing the transformation toward an improved safety culture in healthcare. The dental diagram is intended to standardize the communication and documentation protocol and minimize errors; the Speak-up series is designed to raise awareness, inspire assertiveness, and minimize medical errors. Additionally, the dental diagram could serve as a briefing-and-debriefing tool for the dentist to review with the patient both before and after the surgery to confirm with the patient which tooth should be extracted and to review if everything was satisfactory in the procedure after it is finished. Such a practice would serve as a self-imposed tracking of routine procedures to look for trends such as forgetting to do something in preparation for the surgery, discovering some surprises in the patient's dental history, complications caused, or human errors made. Following is an example of an immediate review of a near-miss that reveals that even with guidance to prevent wrong side or wrong site surgery, mistakes may happen:

> We have quarterly safety rounds in surgery where we focus on the good things that have happened and new procedures. We had a 'great catch' session that featured a neurosurgery case. At the time the great catch occurred, the neurosurgeon was so shaken by what happened that he called in the assistant chief for a consultation in the OR.

The case was to treat a young male with an acute head injury from a skiing accident – he had been brought in by a helicopter. The OR team followed all procedures for right side, right patient – all of the 'rights'. Everyone in the room confirmed the surgical site on the patient's head. The surgeon went out to scrub and came back and when he returned he was shocked to discover that he and the nurses and techs had prepped the wrong side. It was a perceptual thing. We sat down and went through what had happened.

The anaesthesiologist looked at the x-ray and said, I am really surprised that he doesn't have more external trauma – must have had a good helmet. He was looking at the x-ray of the side without the injury. It didn't trigger anyone to detect the mistake. We did our usual root cause analysis – we did it in a week and couldn't do the deep dive into the cognitive aspects of this. There were so many opportunities to disengage from their misperception. The nurse and surgeon put the kid's head in the brace and both made the same perceptual mistake. They talked about how to position the brace. We never had time to explore why they shaved the wrong side of head, and why this didn't this get anyone's attention? Or why the anaesthesiologist's comment didn't trigger a re-look? These are intensely interesting questions. This team followed the procedure for preventing wrong site surgery and still didn't get it right.

This is an example of how challenging it can be to conduct a Root Cause Analysis of sufficient depth to understand the cognitive dimensions of a near miss.

Example 2.2: Landing Overrun by Delta Connection Flight 6448[4]

On February 18, 2007, Delta Connection flight 6448 overran on runway 28 at Cleveland Hopkins International Airport during snow conditions and poor visibility conditions. Three of the seventy-one passengers received minor injuries; the aircraft sustained substantial damage. The NTSB concluded that the probable cause of the accident was 'failure of the flight crew to execute a missed approach when visual cues for the runway were not distinct and identifiable'. Contributing factors were as follows: (1) the crew's decision to descend to the ILS decision height instead of the localizer (glideslope out) minimum descent altitude; (2) the first officer's long landing on a short contaminated runway and the crew's failure to use reverse thrust and braking to their maximum effectiveness; (3) the captain's fatigue, which affected his ability to effectively plan for and monitor the approach and landing; and (4) Shuttle America's failure to administer an attendance policy that permitted flight crewmembers to call in as fatigued without fear of reprisals.

Per the Root Cause Analysis approach, the final outcome is runway overrun. The immediate cause of the overrun is the crew's failure to execute a missed approach – that means failure to elect *not* to land at the intended airport and divert to a different airport or wait for the visibility to improve and try the landing again. Once they committed to land the airplane, the location of wheels touching the runway was about halfway down the runway, which is too far down even without the meteorological challenges. The sloppiness in the decision to land as well as in the execution of the landing is at least partially explained in the third layer of the causal model: why did the airplane overrun? Because the crew failed to execute a missed approach – why did the crew not execute a missed approach? – because they did not think it was a problem – why did they not think it was a problem? – because they made a mistake in their calculations – why did they make a mistake in their calculations? – because the captain was fatigued – why was the captain fatigued? – because he was overworked and was on the tail end of his multi-segment trip. The process could go on a bit further, but once we get into organizational aspects like crew scheduling or corporate fatigue policy, the contributing factors tend to fan out beyond a realistic ability to influence them. The concurrent contributing factor here is the reported deficiencies in co-pilot's flying skills. So, a skill-limited copilot and a fatigued captain tried to land the airplane through poor visibility.

This accident brought to light, again, issues related to pilot proficiency and fatigue. The issue of pilot fatigue is recognized as such a serious challenge that it has made the NTSB's 'Most Wanted' list.

---

4   NTSB Report Number AAR-8/01, April 15, 2008. Available at <http://www.ntsb.gov/publictn/A_Acc1.htm>.

Since fatigue is a complex phenomenon involving physical and cognitive deterioration and is difficult to objectively diagnose, the most common causal association in fatigue cases is sleep deprivation, which in turn is typically associated with duty time limitations. So, there is a certain level of political pressure[5] for the FAA to develop a fatigue policy or to modify the existing duty-time rule (14C.F.R. § 121.471) and for the airlines to enforce duty-time limitations that consider the number of flight legs flown and duty start times, not just the total duty time.

Figure 2.2 presents a partial list of all the NTSB recommendations[6] regarding human fatigue that has made this issue one of NTSB's top priorities. This list can be regarded as a cultural artifact because it is symbolic of the importance that the aviation industry places on addressing human fatigue challenges. It is also symbolic of the unique relationship between the NTSB and the FAA. The NTSB can only make a recommendation; the FAA has to consider this recommendation, but may not want to, or have capacity to, implement the recommendation. The FAA has to consider not only the opinions of those who will be affected (the aviation community) but also has to consider the costs of implementation (to the FAA and to all the stakeholders).

The issue of pilot fatigue, particularly for those crew members commuting from different cities to their 'base,' came back into the spotlight after the Colgan Air crash[7] in February 2009. Soon after this accident, Colgan Air implemented a 'no questions asked' fatigue policy that allowed their pilots and flight attendants to report if they were fatigued and have someone else take their flights. However, this policy may have been either abused or the fatigue conditions are worse than previously known because a January 2010 report (Zremski, 2010) indicates that the airline is tightening its fatigue policy.

The above two cases illustrate a convergent-divergent effect between incidents and solutions. On the convergence side, multiple contributory factors converge and result in the undesirable event. Case analysis of each event leads to multiple levels of solutions – from individual to organizational to national to global level – hence, the term 'divergence'. In the wrong-site extraction case, the contributory factors that converged to create the event were consistent with contributory factors across wrong-site surgeries in general (not limited to dentistry or to that specific procedure), and hence, the industry-wide or divergent advocacy of the Universal Protocol[8] serves as a critical campaign. Similarly, the fatigue issue raised in the landing overrun incident was consistent with several other airliner accidents and

---

5   NTSB Safety Recommendation Dated August 7, 2009. Available at <www.ntsb.gov/recs/letters/2009/A09_61_66.pdf>.

6   Available at <http://www.ntsb.gov/Recs/mostwanted/aviation_reduce_acc_inc_humanfatig.htm>.

7   Preliminary Report is available at <http://www.ntsb.gov/>.

8   Universal Protocol poster available at <http://www.jointcommission.org/standards_information/up.aspx>.

Recommendations & Advocacy

Most Wanted Transportation Safety Improvements Federal Issues

Aviation

Reduce Accidents and Incidents Caused by Human Fatigue in the Aviation Industry

Safety Recommendations

A-94-194 (FAA)
Issued November 30, 1994
Added to the Most Wanted List: 1995
Status: Open—Unacceptable Response
Revise the Federal Aviation Regulations contained in 14 CFR Part 135 to require that pilot flight time accumulated in all company flying conducted after revenue operations—such as training and check flights, ferry flights and repositioning flights—be included in the crewmember's total flight time accrued during revenue operations. (Source: *Commuter Airline Safety*, [NTSB SS-94-02]).

A-95-113 (FAA)
Issued November 14, 1995
Added to the Most Wanted List: 1996
Status: Open—Unacceptable Response
Finalize the review of current flight and duty time regulations and revise the regulations, as necessary, within 1 year to ensure that flight and duty time limitations take into consideration research findings in fatigue and sleep issues. The new regulations should prohibit air carriers from assigning flight crews to flights conducted under 14 CFR Part 91 unless the flight crews meet the flight and duty time limitations of 14 CFR Part 121 or other appropriate regulations. (Source: *Investigation of an Uncontrolled Collision with Terrain, Air Transport International, Douglas DC-8-63, N782AL, Kansas City International Airport, Kansas City, Missouri, February 16, 1995* [NTSB/AAR-95-06]).

A-97-71 (FAA)
Issued September 9, 1997
Added to the Most Wanted List: 1999
Status: Open—Unacceptable Response
Review the issue of personnel fatigue in aviation maintenance; then establish duty time limitations consistent with the current state of scientific knowledge for personnel who perform maintenance on air carrier aircraft. (Source: *The Investigation of the In-flight Fire and Impact with Terrain, ValuJet Airlines Flight 592, DC-9-32, N904VJ, Everglades, near Miami, Florida, May 11, 1996* [NTSB/AAR-97-06]).

A-06-10 (FAA)
Issued February 7, 2006
Added to the Most Wanted List: 2006
Status: Open—Unacceptable Response
Modify and simplify the flight crew hours-of-service regulations to take into consideration factors such as length of duty day, starting time, workload, and other factors shown by recent research, scientific evidence, and current industry experience to affect crew alertness. (Source: *Collision with Trees and Crash Short of the Runway, Corporate Airlines Flight 5966, BAE Systems BAE-J3201, N875JX, Kirksville, Missouri, October 19, 2004* [NTSB/AAR-06-01]).

A-07-30 (FAA)
Issued April 10, 2007
Added to the Most Wanted List: 2007
Status: Open—Acceptable Response
Work with the National Air Traffic Controllers Association to reduce the potential for controller fatigue by revising controller work-scheduling policies and practices to provide rest periods that are long enough for controllers to obtain sufficient restorative sleep and by modifying shift rotations to minimize disrupted sleep patterns, accumulation of sleep debt, and decreased cognitive performance. (Source: Recommendation letter to the FAA regarding four runway incursions and *Attempted Takeoff from Wrong Runway, Comair Flight 5191, Bombardier CL-600-2B19, Lexington, Kentucky, August 27, 2006* [NTSB/AAR-07/05]), April 10, 2007.

A-07-31 (FAA)
Issued April 10, 2007
Added to the Most Wanted List: 2007
Status: Open—Acceptable Response
Develop a fatigue awareness and countermeasures training program for controllers and for personnel who are involved in the scheduling of controllers for operational duty that will address the incidence of fatigue in the controller workforce, causes of fatigue, effects of fatigue on controller performance and safety, and the importance of using personal strategies to minimize fatigue. This training should be provided in a format that promotes retention, and recurrent training should be provided at regular intervals. (Source: Recommendation letter to the FAA regarding four runway incursions and *Attempted Takeoff from Wrong Runway, Comair Flight*

Figure 2.2 Partial list of NTSB recommendations regarding human fatigue

therefore it made the NTSB's Most Wanted list. Again, diverging on the solution side and applying it to multiple organizations.

Additionally, we also want to note the cross-industry effect of divergence. For example, the human fatigue issue raised in the runway overrun example is just as critical in healthcare as it is in aviation. Therefore, we are now starting to see organizations like the American Congress for Obstetricians and Gynecologists (ACOG) release statements that acknowledge the fatigue problem among physicians and recommend ways to minimize fatigue-related errors. We are not suggesting that pilot fatigue issues caused ACOG to acknowledge their own fatigue issues, but the correlation is interesting. ACOG issued the following press release:[9]

> All obstetrician-gynecologists should evaluate the effects that fatigue may have on their ability to care for patients and adjust their workloads, work hours, and time commitments when feasible to avoid fatigue when caring for patients, according to The American College of Obstetricians and Gynecologists (ACOG). In its ongoing commitment to patient safety, ACOG released the new opinion recently in the February issue of Obstetrics and Gynecology. Although there are few published studies on the effects of fatigue on physicians, there is increasing awareness that fatigue, even partial sleep deprivation, impairs performance.

Another aspect of divergence is global impact. For example, the World Health Organization[10] has recognized the importance of preventing medical errors and it is broadening data collection and implementation of interventions to rural and non-hospital settings, which are more common in global healthcare. Again, the link from local to global is not causal, but it is a bit more than coincidental.

Survey Analysis

Survey questionnaires are versatile instruments that could be used to measure key factors across all four levels of the Safety Culture Pyramid as well as to determine the inter-factor influences both within and across the four levels. Further, these characteristics could be used to: (a) compare organizations within the same industry, (b) compare organizations across multiple industries, (c) measure the effects of interventions using pre- and post-intervention comparisons, and (d) conduct regular evaluations over a long period of time to describe longitudinal evolution. However, the survey questionnaires must be carefully designed and their reliability and validity must be tested.

---

9   ACOG press release on fatigue effects available at <http://www.acog.org/from_home/publications/press_releases/nr02-01-08-3.cfm>.

10   2009 WHO patient safety information booklet available at <http://www.who.int/patientsafety/en/>.

Safety performance survey Safety behaviours and outcomes can be tracked by a safety performance survey questionnaire. For example, Vogus and Sutcliffe (2007) report their success with the Safety Organizing Scale (SOS), which is a nine-item measure of self-reported behaviours enabling safety culture. This survey instrument was distributed to 13 hospitals across the country, varying in both size and location (urban/rural). The survey instrument was constructed based on the tenets of High Reliability Organizations and tested for validity and reliability. Vogus and Sutcliffe also collected performance metrics such as average patients per Registered Nurse (RN), number of units, average unit size, reported medication errors, and reported patient falls for the six months after the collection of the survey data. A high score on SOS is considered a positive safety culture. The research reported negative correlation between SOS scores and patient-to-RN ratio, medication errors, and patient falls. Therefore, the higher the SOS score, the lower the medication errors and patient falls.

Chang and Yeh (2004) used a different, but still survey-based, approach to develop an airline safety index that measures safety performance relative to other comparison airlines (data are de-identified). A fuzzy multi-attribute decision-making approach was developed to obtain a safety index for each airline. This approach has four dimensions: management, operations, maintenance, and planning. Under management, the following types of data are collected via fuzzy assessment of survey instruments: safety policy and strategy, management attitude/commitment to safety, and employee attitude/commitment to safety. The *safety policy and strategy* parameter aligns very well with the Safety Strategies in the Safety Culture Pyramid and *the management and employee safety attitudes/ commitment* connects with Safety Climate and Safety Values (the commitment portion). Under operations, competency status of flight crew and compliance with aviation task procedures are collected via the survey. These two parameters align with the Safety Strategies layer of the Safety Culture Pyramid. According to this method, the larger the safety index, the better the safety of the airline. Again, this example simply goes to illustrate that survey questionnaires could be used to collect safety performance data.

Safety climate survey Survey instruments are most commonly used to assess *safety climate* – the employee attitudes and opinions regarding safety. A typical survey instrument consists of several items that may be scored on a Likert-type scale (e.g. indicate agreement with the statement in the item on a scale of one to five, where one equals strongly disagree, two equals disagree, three equals neutral, four equals agree, and five equals strongly agree). Responses to these items are collected and analyzed using either a confirmatory or an exploratory factor analysis technique. A confirmatory factor analysis pre-groups the questionnaire items in accordance with previously tested survey instruments or theoretical constructs; while an exploratory factor analysis lets the statistical analysis software cluster the items and a unifying label is used to identify that cluster. Once the factors are identified, each of them can be tested for reliability and validity.

Flin, Mearns, O'Connor, and Bryden (2000) reviewed 18 industrial safety climate surveys. Their three most strongly emergent themes were management, safety system, and risk. Management plays a critical role in setting policies, providing resources, enforcing safety-oriented behaviours and encouraging communication of hazardous conditions or errors, as well as enforcing discipline against employees who exhibit poor safety behaviours. Flin et al. (2000) noted that all 18 safety climate surveys contained individual items or factors that attempted to measure management's commitment toward safety programmes. The second theme, safety system, is about creating an organizational infrastructure, such as safety committees, that facilitates continuous improvement of safety performance. The employee opinions tend to reflect the effectiveness of such safety systems in improving safety performance. The third theme seeks to determine whether the employees have appropriate perception of risks, the degree to which employees are involved in managing risks, their personality disposition toward risk. Overall, this study indicates that there is support in the literature for safety strategies to influence safety performance.

Scott, Mannion, Davies, and Marshall (2003) conducted a detailed review of nine survey questionnaires that are used in healthcare and four that are used in other industries but could be used in healthcare. All thirteen instruments examined employee 'perceptions and opinions about their working environment (safety climate), but only a few, such as the Competing Values Framework and the Organizational Culture Inventory, try to examine the values and beliefs that inform those views'. Further, 'none of the instruments examined underlying assumptions that guide the attitudes and behaviours and form the stable substrate of culture'. The Competing Values Framework represents a typological approach, characterizing organizational cultures into four types: clannish, hierarchical, market-oriented, or adhocratic. Each organization usually has characteristics of more than one of these types. The Organizational Culture Inventory represents a dimensional approach and describes culture in terms of thirteen dimensions: orientation to customers, orientation to employees, congruence among stakeholders, impact of mission, managerial depth/maturity, decision-making/autonomy, communication/openness, human scale, incentive/motivation, cooporation versus competition, organizational congruence, performance under pressure, theory S versus theory T (Scott et al. 2003). From our perspective, this study supports the notion that values and beliefs (the foundation of our Safety Culture Pyramid) influence attitudes and opinions (the safety climate).

Colla, Bracken, Kinney, and Weeks (2005) conducted a review of patient safety climate surveys and identified nine surveys. They found that the most common areas addressed by the surveys included: leadership, policies and procedures, staffing, communication and reporting. They concluded that the psychometric properties of the surveys varied considerably. They found five surveys to be comprehensive and sufficiently strong psychometrically to recommend them for further use. These surveys include the following: the Veterans Administration Patient Safety Culture Questionnaire (Burr et al. 2002); the Hospital Transfusion

Service Safety Culture Survey (Sorra and Nieva, 2002); the Hospital Survey on Patient Safety (Sorra and Nieva, 2003); and the Safety Attitudes Questionnaire (Sexton and Helmreich, 2004). The Safety Attitudes Questionnaire had been used to investigate the relationship between safety climate and patient outcomes (safety performance). Colla et al. (2005) concluded that more research is needed to establish the linkage between safety climate surveys and patient outcomes. They recommend that until such findings are available, survey results should be interpreted with caution.

Singla, Kitch, Weissman, and Campbell (2006) also reviewed safety climate surveys used in healthcare. They agreed with Colla et al. (2005) in their general conclusions that the survey instruments would benefit from improved psychometric analysis and demonstration of their ability to predict patient outcomes. Singla et al. (2006) studied 13 patient safety instruments that yielded 23 different dimensions, which they grouped into the following general categories: management/supervision, risk, work pressure, competence, rules, and miscellaneous. They found that the content of the surveys varied considerably. The most common dimensions to be included in the surveys were management and institutional commitment to safety, communication openness, and beliefs about the causes of errors and adverse events. Interestingly they found that risk taking was only included in a single survey. Moreover, while professionalism is regarded by many as critical to patient safety in day-to-day healthcare delivery, it was not addressed directly by any of the surveys studied. Singla et al. (2006) point out that only the Safety Attitudes Questionnaire and the AHRQ Hospital Survey on Patient Safety have substantial data sets available for normative comparisons and benchmarking purposes.

Consistent with the analysis of Colla et al. (2005) and Singla et al. (2006), Pronovost and Sexton (2005) state that in healthcare 'the enthusiasm for measuring culture may be outpacing the science' (p. 231). They offer suggestions for bridging the scientist–practitioner gap in assessing safety culture in healthcare. They discuss the importance of using psychometrically established measures such as the Safety Attitudes Questionnaire. Pronovost and Sexton (2005) also contend that it is important to use multidimensional measures to assess safety culture, thus allowing researchers to track the numerous underlying drivers of patient safety. They recount the situation where units with outstanding levels of collaboration, communication, openness, and excellent safety performance may develop a sense of immunity to error and thus fail to recognize the impact of stressors that may lead to medical mistakes. They also point out the importance of assessing safety culture across all units of an organization since they have found considerable variation of safety culture among units within a single organization. Additionally, these authors discuss the importance of feeding back assessment results to staff as well as to members of senior management so that the results can be collectively understood and lead to informed action that improves patient safety.

Sexton et al. (2006) report on their healthcare research with the Safety Attitudes Questionnaire (SAQ). About 25 per cent of the items in the SAQ trace their lineage to commercial aviation and the Flight Management Attitudes

Questionnaire (Helmreich et al. 1993; Helmreich and Merritt 1998), while the remaining items were developed by consulting conceptual models and subject matter experts. The survey was refined by pilot testing and exploratory factor analyses. The final survey consists of six factors that measure teamwork climate, safety climate, perceptions of management, job satisfaction, working conditions, and stress recognition. The entire survey consists of 60 items (answered on five-point Likert scales), demographic items, and a section for open-ended comments. Versions of the SAQ are available for use in intensive care units, operating rooms, general inpatient areas, and ambulatory care. These versions have minor wording changes to reflect reference to the medical units where the surveys are being administered. The Sexton et al. (2006) results are based on 10,843 surveys representing 203 clinical areas collected from healthcare providers in the United States, United Kingdom, and New Zealand. Hospital safety climates were not monolithic and the authors found it important to assess patient safety at the clinical area level where they found more variability between than within clinical areas. The SAQ, norms, and benchmarking data are available at the following website: <http://www.utpatientsafety.org>.

Development of the AHRQ Hospital Survey on Patient Safety Culture (HSOPSC) is documented by Sorra and Nieva (2003, 2004). The AHRQ survey is available at <http://www.ahrq.gov/qual/patientsafetyculture/hospform.pdf>. It consists of nine safety culture dimensions: supervisor/manager expectations and actions promoting safety, organizational learning, communication openness, feedback and communication about error, non-punitive response to error, staffing, hospital management support for patient safety, teamwork across hospital units, hospital handoffs, and transitions. The survey also contains demographic questions and four safety outcome variables: overall perceptions of safety, frequency of event reporting, safety grade, and number of events reported. A comparative database for the HSOPSC is available at the following website: <http://www.ahrq.gov/qual/patientsafetyculture/hospsurvindex.htm>. In 2009 the AHRQ database included responses from over 600 hospitals and 196,000 individual healthcare providers.

Smits, Christiaans-Dingelhoff, Wagner, van der Wal, and Groenewegen (2008) analyzed results from a Dutch translation of the HSOPSC completed by 583 staff at hospitals in the Netherlands. Based on factor analytic techniques, they found a very similar structure to the original questionnaire. They also reported acceptable reliability and good construct validity. They concluded that HSOPSC is a useful instrument to assess patient safety in Dutch hospitals. They also suggest that HSOPSC appears adaptable for translation into other languages to allow for more cross-country comparisons on patient safety.

Blegen, Gearhart, O'Brien, Sehgal, and Alldredge (2009) also report on recent psychometric analyses of the Hospital Survey on Patient Safety Culture. Their sample included 454 healthcare providers from three hospitals located in the United States. The HSOPSC was administered in a pre-post methodology to assess multidisciplinary interventions aimed at improving safety culture. Their results showed moderate-to-strong validity and reliability for HSOPSC. Subscales correlated with perceived outcomes: however, they caution that addition research is needed to evaluate predictive validity.

Safety strategies survey   Safety strategies consist of factors like organizational mission, leadership, structures, policies, procedures, practices, and training. Survey questionnaires could be used to assess the role of leadership in improving the safety culture, effectiveness of policies and procedures, including those used to change the existing policies and procedures, and training effectiveness. Further, if the change strategies are effective, the structures, policies, procedures, and practices will continue to evolve with the changing needs and new knowledge gained from successes as well as failures. Consequently, the organization will become a learning organization. Again, survey questionnaires could be used to evaluate the organizational learning ability.

While it is widely acknowledged that it is the leaders' responsibility to set the tone for the organization, to guide its strategic development, and to serve as role models for rest of the employees, often leaders are challenged in implementing their strategies. Guth and MacMillan (1986) present results of a qualitative study that builds a simple yet profound model describing why senior management's strategy might fail. Senior management's change strategy might fail if the middle management is not convinced that: (a) a change is necessary and timely, (b) the proposed strategy is the right strategy to achieve the intended goal, and (c) their self-interest is aligned with the corporate goal. Survey instruments may be used to gauge the level of middle-management support in accordance with the above three parameters.

Once the senior managers and middle managers are fully aligned in their goals, strategies, and respective self-interests, successful implementation of the strategy depends on how well management attitudes can translate into behavioural intent and how well the intent manifests into actual behaviour. Again, management attitudes, behavioural intent, and actual behaviours could be assessed with survey questionnaires. Rundmo and Hale (2003) analyzed the relationship between managers' safety attitudes, behavioural intentions, and their self-reported behaviours. They identified ten belief dimensions of management attitude: management safety commitment and involvement; fatalism concerning accident prevention; management attitude toward rule violation; management safety talk and risk communication with the employees; personal worry and emotion about safety issues; powerlessness (fatalism) to do anything about safety problems; priorities of safety in general; mastery of own work situation related to safety; perceived hindrances for safety involvement, and management awareness about potentially-hazardous risks. Behavioural intent was measured across four dimensions: motivation of employees and monitoring of safety practices, time spent on improving procedures and safety regulations, management involvement in design and development of equipment, and involvement in safety instructions and training. To get a sense of the gap between behavioural intent and actual behaviours, managers were asked how much time they actually spent versus the amount they needed to spend on eight job-related activities during their work hours: operational issues, productivity, product quality, job satisfaction, conflict resolution, safe working environment, and environmental issues. Rundmo and Hale report that their sample of managers spent too little time on operational issues and they should spend less time on all other issues and more on operational issues. The analysis of relationship between attitudes, intentions, and behaviours revealed that 'attitude dimensions exerting the most significant influence were *management commitment and involvement in safety work*, which influenced behavioural intentions with regard to *motivation and monitoring* as well as *procedures and safety regulations*'. Again, there is support in the literature for influence between leadership (safety strategy) and safety performance through safety attitudes (safety climate).

There are a variety of change strategies that could be considered, depending on the safety goals of the organization. While most strategies aim at changing existing organizational structures, policies, procedures, or practices, the choice of a particular strategy should be preceded by assessment of the safety climate and safety performance. These two assessments will form a baseline for pre/post-intervention comparison. If the safety performance assessment leads to the suspicion that latent systemic errors may be present, an error and hazard reporting system could be implemented. Survey instruments could be developed to measure the effectiveness of such a system. A simple tracking of the number of reports submitted per month and the number and type of changes made in response to those reports will help demonstrate how the reporting system, coupled with the subsequent investigation and remediation, is gradually improving the organization, making it a learning organization. Such ability of an organization can be measured along seven dimensions using a Dimensions of Learning Organization Questionnaire (DLOQ): 'Continuous learning, inquiry and dialogue, collaboration and team learning, systems to capture learning, empower people, connect the organization, and provide strategic leadership for learning' (Marsik and Watkins, 2003).

Safety values survey In 1984, Geert Hofstede conducted one of the earliest, and perhaps the largest, cross-cultural research programmes ever in a systematic study of work-related values across more than 50 countries. He found that within a given industry, certain national differences are seen in hierarchical differences and social distance (he called it power distance), in preferences for individualism or collectivism, and in tolerance for uncertainty (he called it uncertainty avoidance). Later, Helmreich and Merritt (1998) showed the effects of differences in national culture among airline pilots and surgeons. Their results act to confirm the theory and prior findings of Hofstede (1984). Taylor and Patankar (1999) found that some of these differences in national culture reported by Hofstede, and confirmed by Helmreich and Merritt, also affect airline mechanics and their managers. Based on these studies, one could say that professionals from Asian countries tend to be more collectivistic than those from Western European countries. When

Patankar and Taylor (2004) measured such differences across the United States, they discovered that maintenance professionals from the East or the West coast of the United States were more individualistic than those from the Mid-western or Central regions. Thus, the notion of 'national differences' could be applied to distinctive geographic locations within the same country. From a professional perspective, it has been demonstrated that certain professionals such as pilots are more individualistic than surgeons (Helmreich and Merritt, 1998) and aircraft mechanics are more individualistic than pilots (Patankar and Taylor, 2004). All these cross-cultural studies focused on deeply held personal values regarding work and demonstrated that differences in work-related values yield discernable patterns of work habits or behaviours. Thus, these studies illustrate the relationship between values and behaviours.

Influence of Safety Values, Safety Strategies, and Safety Climate on Safety Performance

Kouabenan (1998) demonstrates how personal values such as fatalistic beliefs and national cultural norms like mystic practices can influence the perception of accidents and consequently create a tendency 'to take more risks and neglect safety concerns'. A twenty-item survey questionnaire was used to measure responses along two scales: personal belief in fate and risk-taking tendency. Most interestingly, professional drivers were the most fatalistic (and took the greatest risks) and engineers were the least fatalistic. So, there's some support to the notion that personal values and beliefs could impact safety performance – if the personal belief is fatalistic, even placement of legal barriers to prevent over-speeding in one's car may not dampen the individual's risk-taking tendency, and therefore there appears to be a link between safety values and safety behaviours somewhat independent of the mitigating strategies.

Johnson (2007) established that safety climate surveys can provide reliable assessment of the prevailing psychosocial conditions and organizational conditions; further these conditions have been known to impact the safety-oriented behaviours of employees (Neal and Griffin, 2002; Cooper and Phillips, 2004). In addition to the specific psychosocial aspects of the extant safety climate, Arezes and Miguel (2003) note that there are a wide variety of performance factors such as lost time injuries, number of accidents/incidents, and damage events that could be used as indicators of safety. While a decline in such numbers may indicate that the safety of the operation is increasing, there is no guarantee that the trend will continue. Also, there may be an implicit incentive to under-report incidents in order to maintain the positive image of the organization. Nonetheless, performance measures can be correlated with attitudinal changes that are in turn related to specific safety culture change efforts (Patankar and Taylor, 2004) and subsequently used to demonstrate the financial benefits of the change programme. Patankar and Taylor (2004, Ch.8) demonstrate that both positive as well as negative financial returns on investment are possible, depending on how closely the content matches the intent.

Fernández-Muñiz, Montes-Peón, and Vázquez-Ordás (2007) provide a comprehensive example linking most aspects of the Safety Culture Pyramid. They developed a safety culture model consisting of: (a) Safety Management System, (b) Management Commitment, (c) Employee Involvement, and (d) Safety Performance. Their model is very similar to the Safety Culture Pyramid and so elements of this model could be used to demonstrate the application of the Safety Culture Pyramid. For example, the Safety Management System component is essentially Safety Strategies in the Safety Culture Pyramid; parts of the Management Commitment component are more aligned with the Safety Values concept in the Safety Culture Pyramid and some parts are better aligned with Safety Climate and Safety Performance; Employee Involvement is partly a Safety Strategy because it is a way to build buy-in among the employee group and it is partly Safety Performance because the employees will be expected to behave differently as they get involved in the implementation of the Safety Management System; finally, Safety Performance is the same in both models. Fernández-Muñiz et al. tested their model across 62,146 companies (mostly small companies) across Spain, including construction, industrial, and service sectors. The results of their study confirm that management commitment has a direct, positive, and statistically significant influence on both employees' involvement and safety management systems; safety management systems and employee involvement have significant positive influence on safety performance. Again, there is support for the Safety Culture Pyramid Model and the related notion that safety performance is influenced by safety climate, safety strategies, and safety values.

Assessment Across Professional Groups, Organizations, and Industries

Safety Climate surveys can be used across professional groups within the same industry (e.g. Patankar, 2003); across multiple organizations within the same industry (e.g. AHRQ Survey; Safety Attitudes Questionnaire; Helmreich and Merritt, 1998, etc.); across multiple industries (e.g. Safety Attitudes Questionnaire; Harvey, Erdos, Jackson, and Dennison, 2004; Fernández-Muñiz et al. 2007, etc.). However, since the factors that define safety climate have not been completely delineated or systematically agreed among the researchers, we don't quite have industry or national-level benchmarks. Nonetheless, the use of safety climate surveys is stabilizing and the component factors are being tested more rigorously not only for their validity and reliability, but also for their influences on each other.

There are two other ways to use safety climate questionnaires to further test and stabilize the underlying model: longitudinal assessment and pre- and post-intervention assessment. Longitudinal assessment is important because without such assessment, a safety climate survey is limited to one snapshot of the attitudes and opinions at that time for the sample population. Without continued, regular administration of the surveys, it is impossible to determine trends and develop a more complete understanding of the trends.

Karl Weick's message, 'the environment that the organization worries about is put there by the organization' (Weick, 1979, p. 152) is particularly salient. We want to emphasize this quote because we believe the message here is that for a safety *culture* to change, the organization must be equally willing to change its structures, processes, and policies. This is particularly important when employees start using a newly established error-reporting system or a safety management system – they expect that structural or procedural changes will be made in response to their participation, and if the requisite changes are not forthcoming, they will lose interest in the process and therefore in the organization and/or the leadership. With respect to Maintenance Resource Management training, which was intended to raise the awareness regarding human factors issues in aviation maintenance, Taylor and Christensen (1998) note the following observation: 'Frustrated with slow progress in achieving the promise of MRM training, a sizable number of AMTs [aviation maintenance technicians] saw a greater need to speak up – perhaps even in anger or frustration – as the only path for improvement' (p. 161). The MRM training raised the expectations among the employees, but did not deliver the responsive changes promptly enough.

In summary, survey instruments can be used to measure safety performance, to describe the safety climate, to determine the effectiveness of safety strategies or the level of support for certain strategies, and to gauge the alignment of safety values among employees and management. Since surveys tend to engage the employees through solicitation of feedback, the employees tend to expect that something will change in response to the findings from the survey. If nothing changes, the employees will lose faith in the company/management and will be less enthusiastic about organizational changes in the future.

Qualitative Analysis

A comprehensive analysis of safety culture needs qualitative methods such as field observations, artifact analysis, interviews, focus group discussions, and dialogue with key individuals in the organization (e.g. Maxwell, 2004). Typically, a safety culture project may start with interviews, focus group discussions and field observation. In the early stages of the research, interviews are essentially orientation sessions to get to know the industry, the organization, and the facility. Interviews then lead to focus group discussions to elicit key issues or unique interests/characteristics of the organization/location. Concurrent field observations allow the researchers to better understand the work environment of their subjects. Often interviews and/or focus group discussions may transition to a field trip across the organization's various facilities. Also, during these meetings and tours, it would be a good idea to collect cultural artefacts such as newsletters, programme stickers, safety culture brochures or posters, organizational charts, employee evaluation forms, etc. Occasionally, the interviews/focus groups turn into deep dialogue with key informants in the organization. These are precious opportunities to listen and learn all about the history of the organization, the past stories, legends and heroes from the organization and the overall sense of anxiety or comfort among the employees. This section provides more details about each qualitative data collection technique and provides examples where appropriate.

Field observations Field observation may include naturalistic or participant observation. In naturalistic observation the goal of the research is to observe participants in their natural setting and avoid any intervention or interference with the normal course of events. Participant observation is the primary research approach of cultural anthropology. Here the researcher develops an intensive relationship with the participants by typically interacting over an extended period of time. Ethnography represents a particular approach to field observation by describing human culture from a holistic perspective through direct but non-intrusive observation. This approach is described in the writings of Hammersley and Atkinson (1983), Erickson and Stull (1997), and Spradley (1979).

From a safety culture perspective, Line Oriented Safety Audit (LOSA) is a popular normal operations observation protocol which is used in both aviation and healthcare (cf. Thomas, Sexton, and Helmreich, 2004). Instead of using external observers, as in a traditional ethnographic study, LOSA protocol trains peer observers to observe the flight/maintenance crew or the surgical team in their natural environment under normal operations. This is not an examination/evaluation. The peer observers are specially trained in accordance with the Threat and Error Recognition Model (Helmreich, Klinect, and Wilhelm, 1999) to mark all the threats that the subjects faced, their errors, and their successful recoveries. These data are then used to provide constructive feedback to the teams and improve their safety performance. The teams are always free to decline or end a LOSA-type observation. The theory behind the LOSA-type observations and prompt feedback is that professionals in high-consequence industries routinely manage threats and errors. Some teams develop superior abilities to overcome systemic limitations or deficiencies in one another without compromising their mission, while other teams crumble under comparatively benign threats. The LOSA-type observations enable the teams to learn about their strengths and their weaknesses promptly and, most importantly, without having to wait for an accident. Organization-wide use and documentation of learning through LOSA-type observations would be a mark of a continuously learning organization.

Artifact analysis Cultural anthropologists have developed sophisticated and detailed techniques of describing cultures based on their artefacts. While most of the artefacts are in the form of physical articles that were used by the community or symbols, drawings, and paintings created by the community, some artefacts are also in the form of stories that convey core values and key experiences so that the future generations learn from the past (Bernard and Spencer, 1996). In organizational safety culture, there are similar artefacts.

Artefacts are an important element of the overall cultural assessment because they are created by the people who are part of the organization, and the design of such artefacts is therefore reflective of the organization's cultural evolution. According to anthropologists, artefacts include physical evidence as well as key stories, legends, and myths that are passed down through generations because deeply held cultural values tend to be passed across generations through stories. Examples of such artefacts, in safety culture, include the following:

- Company publications – mission statement, safety policies, safety performance reports, safety training manuals, accident/incident reporting forms, error-reporting forms, policy or procedure change protocols, accident/incident reports, posters, and warning/caution signs and so on.
- Personal protective safety equipment actually used by the employees – such as goggles, masks, ear protection equipment, fluorescent vests, hard hats, steel-toe shoes.
- Safety markings in the facility – symbols, words, and demarcation lines clearly identifying hazards or hazardous areas as well as required safety equipment (hard hats, goggles, masks, etc.).
- Employee training and certification programmes – including safety awareness training, specific behavioural training to reduce errors and injuries, first-responder type training, incorporation of safe practices in technical training, human performance and team performance issues, system safety programmes (e.g. Crew Resource Management and Maintenance Resource Management).
- Symbols of safety initiatives – logos, stickers, t-shirts, lanyards, awards/medals, certificates.
- Hazard or error reporting programmes – specific non-punitive programmes to encourage all employees to self report their errors and report specific hazards (e.g. Aviation Safety Action Program, Aviation Safety Reporting System, Patient Safety Reporting System, MEDMARX Database).
- Peer observation programmes – programmes designed to raise the awareness of safety threats during normal operation and proactive, non-punitive analysis of ability to mitigate such threats (e.g. Line Operations Safety Audit).
- Stories of past experiences – what makes heroes and legends in the organization? What happened when someone pointed out a serious safety issue – was the individual punished, was the safety issue resolved, is there a general sense of positive achievement or is there a bitterness that signifies distrust? Are safety champions valued or are they viewed as barriers to productivity? How have critical safety challenges been addressed when they threatened to compromise productivity? Are there any well-recognized safety champions – on the employee side (including labour union representatives) as well as the management side? What are the senior employees telling the junior employees?
- Newsletters, methods and tools used to communicate safety practices; report, analyze, and proactively solve safety problems.

Southwest Airlines has published a Safety and Security Commitment[11] statement for its employees and passengers. Johns Hopkins Hospital has a similar statement of commitment, but it emphasizes safety and quality improvement.[12] Both these statements are examples of important cultural artefacts. A comparison of the two statements of commitment indicates that in healthcare, safety and quality are addressed together, while in aviation, they are treated separately. So, strategically, healthcare organizations would be more comfortable applying quality management tools to improve safety, but aviation organizations may be resistive to the use of quality management tools in safety. Further, the aviation industry has made a pledge to not compete on safety. Consequently, industry groups freely share safety data across corporate boundaries. In healthcare, since safety and quality are more tightly coupled, safety, like quality, may be regarded as a competitive advantage. Such subtle, yet critical, distinction between the cultures can be discovered through artifact analysis.

Another example of parallel artefacts in aviation and healthcare is a checklist. Pilots use checklists as part of their routine flying – there's a section for every phase of flight and also for emergencies. Surgeons have started using similar checklists relatively recently. Gawande (2011, pp. 197–200) presents some comparative examples of checklists from aviation and healthcare and argues in strong support of the use of checklists in surgery.

Interviews and focus group discussion Typically, interviews and focus group discussions have been used by safety researchers to develop a basic understanding of the vocabulary, norms, key safety challenges, organizational policies, and procedures. Systematic procedures and best practices can be used to design, facilitate, and analyze information from focus groups and interviews (Krueger and Casey, 2008; Schwarz, 2002; Stewart, Shamdasani, and Rook, 2006).

Such qualitative data are also useful in developing customized survey instruments that can be distributed to a larger audience to collect data on attitudes and opinions. Additionally, stories collected from interviews and group discussions can reveal heroes and legends that symbolize key cultural experiences. These stories tend to reveal deeply rooted meaning and help expose underlying values and assumptions.

Dialogue Dialogue is an intense conversational technique that seeks to address serious topics, underlying values, and unquestioned assumptions. Dialogic communication requires trust, openness, and respect between the participants. Dialogue can offer significant insight into important values and assumptions that are at the foundation of an existing safety culture.

---

11   Southwest Airlines' Safety and Security Commitment Statement available at <http://www.southwest.com/about_swa/safety_commitment.html>.

12   John Hopkins' Safety and Quality Improvement Statement available at <http://www.hopkinsmedicine.org/quality/safety/commitment.html>.

Isaacs (1993) defines dialogue as 'a sustained collective inquiry into the processes, assumptions, and certainties that compose everyday experience' (p. 25). Its primary purpose is to help people engage in productive thinking together. Isaacs regards dialogue as a powerful tool for team learning since it focuses attention on aspects of experience that we frequently ignore and examines hidden assumptions behind conventional problems and ways of thinking. Isaacs' *Dialogue and the Art of Thinking Together* (1999), and Senge's *Fifth Discipline* (1990, 2006) and *Fieldbook* (Senge, Kleiner, Roberts, Ross, and Smith 1994) offer useful guidance for building practical dialogue skills to improve team functioning and organizational learning. Levine (1994) contends that selfless collective listening and team dialogue are learnable and replicable skills that can be used in diverse organizational settings to improve insight, innovative thinking, and performance.

Schein (1993) maintains that dialogue 'is a central element of any model of organizational transformation' (p. 27) and is essential to promote understanding of an organization's culture and constituent subcultures. Dialogue is regarded as a basic skill to recognize tacit assumptions, solve problems, resolve conflict, and learn from experience in complex group settings. 'The process of communicating across hierarchical levels ... will require further dialogue because ... different strata operate with different assumptions' (p. 37). Schein contends that it is necessary to understand ourselves first before we can truly understand others. Recognizing our own limitations can be threatening and cause defensiveness. For dialogue to be effective it must occur in a safe environment where emotional issues can be handled effectively 'without anyone getting burned or burning up' (p. 35). Schein's 1999 text on the consultation process provides a masterful description of dialogue's use to uncover hidden forces and processes in organizational settings.

Abramovitch and Schwartz (1996) reference the ideas of Martin Buber's (e.g. 1947; 2002) philosophy of personal dialogue to explain the personal (I-Thou) and impersonal (I-It) relationships with patients that can occur in the practice of medicine. The authors argue that the failure of physicians to maintain a genuine dialogue (i.e. I-Thou) with their patients has resulted in a 'humanistic crisis' in medicine. They maintain that treating patients as objects (i.e. I-It) results in numerous undesirable effects, such as patient resentment, dissatisfaction, non-compliance with medical directions, malpractice litigation, and overmedication. These authors argue that these negative outcomes can be minimized through the process of authentic physician-patient dialogue as a basis for shared medical decision-making.

April (1999) asserts the centrality of dialogue to organizational change, which occurs in the context of social interaction. Dialogue builds critical thinking and results in insight and a call to action. Stubbs (1998), like Senge (1990), also regards dialogue as fundamental to bringing about transformative change. Dialogue helps achieve this by going beyond typical discussion to tackle 'undiscussables'. If unquestioned assumptions and other undiscussables are not addressed honest communication is prevented, which then blocks significant change. These authors maintain that transformational leaders require dialogue to be successful.

Gerard and Teurfs (1997) succinctly describe the core elements of effective organizational dialogue. These key dialogue skills include: suspending judgment; identifying assumptions; active listening; and inquiring and reflecting. They contend that by applying these skills, dialogue builds community and changes culture through: behavioural transformation (changing how participants relate to one another), experiential transformation (inducing a feeling of community), and attitudinal transformation (creating attitudes that support collaboration and partnership, which improve decision-making).

Bodily and Allen (1999) explain a six step process in developing strategy that relies on dialogue between decision makers and strategy developers to optimize insight and creativity. The authors maintain that dialogue can be effectively used: 'To identify issues and challenges; to develop creative, doable alternatives; to treat economic value and risk meaningfully; to balance the process of creating strategies and the process of evaluating and choosing alternatives; to reason logically and correctly; and to commit to action' (p. 28).

Keatings, Martin, McCallum, and Lewis (2006) describe the use of dialogue in the process of disclosure about the systemic issues of medical error that led to the death of 11-year-old Claire Lewis in 2001. Initial silence from the hospital regarding the medical error and subsequent editing of information by legal advisors resulted in a report that produced feelings of distrust and anger in the child's family. Greater alienation resulted from a meeting with hospital representatives where the family felt their concerns were met with defensiveness, intimidation, and insincerity. Eventually the hospital acknowledged that the death had been preventable and offered an apology. According to the child's father: 'The disclosure and apology melted horrific feelings away, slowly opening the door to a meaningful dialogue between the family and the hospital staff' (p. 1085). As part of the corrective changes by the hospital a patient advocate role was created and filled by the child's father. He is described as bringing 'unique passion, insight, and empathy' to the position. Dialogue that discloses the facts and takes responsibility has led to trust and healing among families impacted by medical errors.

Mazor et al. (2004), using survey methodology, studied the reactions of health plan members to scenarios about medical error with different disclosure dialogues between physician and patient. They found that full disclosure dialogues by physicians that provided an explanation, accepted responsibility for the error and offered an apology were superior to nondisclosure dialogues, which merely expressed regret for the error without accepting responsibility or providing an apology. These full disclosure dialogues showed higher levels of patient satisfaction, trust, and positive emotions and reduced the intention of changing physicians. In some situations full disclosure also reduced the intent to seek legal advice regarding the medical error, although the authors conclude full disclosure dialogue may not always prevent litigation.

Triola (2006) discussed the role of dialogue and other interpersonal abilities in skilled communication among critical care nurses in order to develop a healthy work environment and improve patient safety. She references research by

Maxfield et al. (2005) and Patterson et al. (2002), who identified seven difficult but critical types of conversation that occurred in healthcare, including: broken rules, mistakes, lack of support, incompetence, poor teamwork, disrespect, and micromanagement. Triola suggest several tools and strategies to address these difficult conversational topics, such as: examining root cause analyses and failure mode effects analyses for communication problems, self assessments to measure confrontation skills, best practices to develop open dialogue that is candid and respectful, and communication coaching.

Quasi experimental analysis Applied field research does not have the same ability to manipulate, randomize and control variables as experimental laboratory research. However, effective quasi-experimental designs (Campbell and Stanley, 1963) are available for use in field settings to study the linkage between variables. Safety culture researchers have called for better empirical evidence to test hypothesized relationships between safety climate, safety culture, safety behaviour, and organizational performance (e.g. Pidgeon, 1998; van den Berg and Wilderom, 2004). Wiegmann et al. (2002) suggest that organizational psychology is particularly attuned to the importance of testing causal relationships due to its interest in implementing change interventions to improve safety culture.

A multi-method, quasi-experimental analysis, as suggested by us, will allow for both qualitative and quantitative methods pre- and post-intervention. The shift in safety cultural state will need to be documented with the appropriate tools and we need to bear in mind that there will be factors other than the interventions that may influence the change in climate or culture.

Safety Culture Transformation

Safety Culture Transformation is 'transformational' because it is about changes in values of organizations and its people. Rochon (1998, Ch. 3) discusses cultural transformation in terms of three levels of change in values: value conversion, value creation, and value connection. If we are working with people and organizations that currently value a blame-oriented culture because they believe in punishment, then the transformation process is about getting them to see the need to change their values and be open to non-punitive ways of improving safety performance. If organizational learning is not a stated value, then we would be striving to create it by demonstrating the importance of organizational learning to the overall success of the organization. Similarly, if flight safety or patient safety is not an explicitly stated value, but service quality is, a connection could be made between service quality and flight/patient safety.

Safety Culture Transformation is about shifting the equilibrium of the Safety Culture Pyramid from its current state to a more desirable state. So, we need to fully understand the current state of the safety culture. This process will enable us to understand the values that are being enacted at this time, the strategies that are in place and in balance with the enacted values, current attitudes and opinions of the employees and management, and finally the current behavioural trends and patterns. All the safety culture assessment methods discussed earlier in this chapter may be used to develop a thorough description of the existing safety culture. Then a parallel set of behaviours, attitudes, strategies, and values needs to be developed as the desired cultural state. This approach is consistent with our Safety Culture Pyramid Model and with Sloat's (1996) recommendation to first focus on values and unquestioned assumptions and then make changes to artefacts (he includes organizational structures and processes as artefacts), including employee evaluation and recognition systems and performance standards. Gordon, Moylan, and Stastny (2004) and Shackford (2005) focus on behavioural aspects of cultural change, but they too acknowledge that such changes should be consistent with the organizational values.

So, first we have to thoroughly understand our current values and unquestioned assumptions. Next, we need to ask if anything in our desired cultural state is in conflict with the currently enacted values. For example, if we are trying to get an organization from blame culture to reporting culture, we need the employee behaviours to change (they need to start submitting the reports), we need management behaviours to change (they need to stop disciplining every time someone confesses to making a safety mistake and to start correcting systemic failures), and we need the regulator's behaviour to change (they should not fine or otherwise violate the individual's professional certification). In order for people to make such fundamental shifts, they should be ready to accept changes in the corporate values, they should believe that this change is essential, and they should see that they have nothing to lose and much to gain in this process. The actual implementation of the transformation will require several discussions with the key stakeholders, assurances of being understanding and working with individual self-interests, and delivering on the promises made in this process.

Finally, cultural transformation takes time and effort from all. It needs to be an organization-wide commitment. A number of organizational processes and policies – particularly the evaluation and recognition systems – might have to change, it may cost more than originally envisioned or budgeted, and it is not going to be without struggle. In some cases, certain unwilling employees will need to retire or quit before progress can be made. Labour unions, corporate attorneys, and regulators will all need to be committed to making the transformation. Regular measurements through surveys and performance data should be shared broadly to keep everyone informed of the transformation status.

Chapter Summary

In this chapter, we discussed a variety of tools and techniques to assess different aspects of safety culture. In accordance with the Safety Culture Pyramid model, a blend of qualitative and quantitative methods needs to be used to develop a comprehensive understanding of safety culture. There is evidence in the literature that supports linking safety values, safety strategies, safety climate, and safety performance. Several studies have also demonstrated an influential link between safety values and safety behaviours, safety strategies and safety behaviours, safety climate and safety behaviours as well as from safety values through safety climate to safety performance.

Typically, a study of safety culture starts with qualitative analysis, using interviews, focus group discussions, and dialogue. Then, the study moves on to a review of artefacts such as organizational charts, unique products, awards, safety reports, etc. that signify important developments in the organization's culture. In parallel, we need to start having dialogue with key informants (in anthropological parlance, people who have deep tribal knowledge about the culture) to develop a fuller understanding of culture, and to start collecting specific safety performance cases that illustrate the cultural challenges at the organization. The analysis of such qualitative data will help prepare a survey instrument that could be used to measure safety climate along valid and reliable scales.

The safety climate survey could be used to identify the extant *state* of safety culture in the given organization. A number of different strategies could be used to move the safety culture toward a more desirable state. The literature supports the notion that deep-seated safety values influence safety strategies, which in turn shape safety attitudes; and ultimately, safety climate/attitudes shape behaviours.

In order to transform a given culture, the people in that culture must change. For the people to change, top management must plan the implementation of the new culture very carefully. They must explain to middle management that improving safety culture is the right thing to do; and, that it is in the interest of the organization's survival. Next, they must get the employees to support the initiative by aligning the employee evaluations and reward structure so that the desirable behaviours are rewarded and undesirable behaviours are curbed. Pre- and post-intervention surveys might help illustrate effect of the intervention.

Dialogue is a special technique of engaging individuals in deep, serious discussions aimed at revealing underlying values and unquestioned assumptions. These discussions are not likely to occur unless there is sufficient trust and openness between the interviewer and the participants. Once established, dialogue can be an extremely powerful mechanism to better understand the values and assumptions that are at the foundation of the existing safety culture.

References

Abramovitch, H. and Schwartz, E. (1996). Three stages of medical dialogue. *Theoretical Medicine*, 17(2), 175–187.

April, K.A. (1999). Leading through communication, conversation and dialogue. *Leadership and Organization Development Journal*, 20(5), 231–242.

Arezes, P. and Miguel, A. (2003). The role of safety culture in safety performance measurement. *Measuring Business Excellence*, 7(4), 20–29.

Bernard, A. and Spencer, J. (1996). *Encyclopedia of Social and Cultural Anthropology*. London: Routledge.
Blegen, M., Gearhart, S., O'Brien, R., Sehgal, N. and Alldredge, B. (2009). AHRQ's hospital survey on patient safety culture: psychometric analyses. *Journal of Patient Safety*, 5(3), 139–144.
Bodily, S.E. and Allen, M.S. (1999). A dialogue process for choosing value-creating strategies. *Interfaces*, 29(6), 16–28.
Buber, M. (1947; 2002). *Between Man and Man*. New York: Routledge.
Burr, M., Sorra, J. and Nieva, V. (2002). Analysis of the Veterans Administration (VA) National Center for Patient Safety (NCPS) FY 2000 patient safety questionnaire. Technical report. Rockville: Westat.
Campbell, D. and Stanley, J. (1963). *Experimental and Quasi-experimental Designs for Research*. Chicago: Rand McNally.
Chang, Y. and Yeh, C. (2004). A new airline safety index. *Transportation Research Part B: Methodological*, 38(4), 369–383.
Colla, J.B., Bracken, A.C., Kinney, L.M. and Weeks, W.B. (2005). Measuring patient safety climate: a review of surveys. *Quality and Safety in Healthcare*, 14, 364–366.
Cooper, M.D. and Phillips, R.A. (2004). Exploratory analysis of the safety climate and safety behavior relationship. *Journal of Safety Research*, 35(5), 497–512.
Di Pofi, J. (2002). Organizational diagnostics: integrating qualitative and quantitative methodology. *Journal of Organizational Change Management*, 15(2), 156–168.
Erickson, K. and Stull, D. (1997). *Doing Team Ethnography: Warnings and Advice*. Thousand Oaks: Sage.
Fernández-Muñiz, B., Montes-Peón, J. and Vázquez-Ordás, C. (2007). Safety culture: analysis of the causal relationships between its key dimensions. *Journal of Safety Research*, 38, 627–641.
Flin, R., Mearns, K., O'Connor, P. and Bryden, R. (2000). Measuring safety climate: identifying the common features. *Safety Science*, 34, 177–192.
Gawande, A. (2011). *The Checklist Manifesto: How to Get Things Right*. New York: Picador.
Gerard, G. and Teurfs, L. (1997). Dialogue and transformation. *Executive Excellence*, 14(8), 16.
Gordon, L. (1994). *Gordon's Guide to the Surgical Morbidity and Mortality Conference*. Philadelphia: Hanley and Belfus.
Gordon, R., Moylan, P. and Stastny, P. (2004, October–December). Improving safety through a just culture: Flight Ops/ATC Ops safety information sharing. *Journal of ATC*, 34–38.
Guth, W. and MacMillan, I. (1986). Strategy implementation versus middle management self-interest. *Strategic Management Journal*, July/August, 313–327.
Hammersley, M. and Atkinson, P. (1983). *Ethnography: Principles and Practice*. London: Tavistock Books.

Harvey, J., Erdos, G., Jackson, H. and Dennison, S. (2004). Is safety culture in differing organizations the same thing? A comparison of safety culture measures in three organizations. *Risk, Decision and Policy*, 9(4), 337–346.

Helmreich, R., Klinect, J. and Wilhelm, J. (1999). Models of threat, error, and CRM in flight operations. In: *Proceedings of the Tenth International Symposium on Aviation Psychology* (pp. 677–682). Columbus: The Ohio State University.

Helmreich, R. and Merritt, A. (1998). *Culture at Work in Aviation and Medicine: National, Organizational and Professional Influences*. Aldershot: Ashgate.

Helmreich, R., Merritt, A., Sherman, P., Gregorich, S. and Wiener, E. (1993). The flight management attitudes questionnaire (FMAQ). NASA/UT/FAA technical report 93–4. Austin: University of Texas Press.

Hofstede, G. (1984). *Culture's Consequences: International Differences in Work-related Values* (Abridged Edition). Beverly Hills: Sage.

Isaacs, W.N. (1993). Taking flight: dialogue, collective thinking, and organizational learning. *Organizational Dynamics*, 22(2), 24–39.

Isaacs, W.N. (1999). *Dialogue and the Art of Thinking Together*. New York: Currency Doubleday.

Johnson, S. (2007). The predictive validity of safety climate. *Journal of Safety Research*, 38, 511–521.

Keatings, M., Martin, M., McCallum, A. and Lewis, J. (2006). Medical errors: understanding the parent's perspective. *Pediatric Clinics of North America*, 53, 1079–1089.

Kouabenan, D. (1998). Beliefs and the perception of risks and accidents. *Risk Analysis*, 18(3), 243–52.

Krueger, R.A. and Casey, M.A. (2008). *Focus Groups: A Practical Guide for Applied Research*. Thousand Oaks: Sage.

Levine, L. (1994). Listening with spirit and the art of team dialogue. *Journal of Organizational Change Management*, 7(1), 61–73.

Marsick, V. and Watkins, K. (2003). Demonstrating the value of an organization's learning culture: dimensions of the learning organization questionnaire. *Advances in Developing Human Resources*, 5(2), 132–151.

Maxfield, D., Grenny, J., McMillan, R., Patterson, K. and Switzler, A. (2005). *Silence Kills: The 7 Crucial Conversations in Healthcare*. Provo: VitalSmarts.

Maxwell, J.A. (2004). *Qualitative Research Design: An Interactive Approach* (second edition). Thousand Oaks: Sage Publications.

Mazor, K.M., Simon, S.R., Yood, R.A., Martinson, B.C., Gunter, M.J., Reed, G.W. and Gurwitz, J.H. (2004). Health plan members' views about disclosure of medical errors. *Annals of Internal Medicine*, 140, 409–418.

NASA (2006). NASA procedural requirements for mishap and close call reporting, investigating, and recordkeeping. Retrieved from <http://nodis3.gsfc.nasa.gov/npg_img/N_PR_8621_001B_/N_PR_8621_001B_.pdf> on April 4, 2009.

Neal, A. and Griffin, M. (2002). Safety climate and safety behaviour. *Australian Journal of Management*, 27, 67–75.

NTSB (2002). National transportation safety board aviation investigation manual major team investigations. Retrieved from <http://www.ntsb.gov/Aviation/Manuals/MajorInvestigationsManual.pdf> on April 3, 2009.

Patankar, M.S. (2003). A study of safety culture at an aviation organization. *International Journal of Applied Aviation Studies*, 3(2), 243–258. Okalahoma City: FAA Academy.

Patankar, M.S. and Taylor, J.C. (2004). *Risk Management and Error Reduction in Aviation Maintenance*. Aldershot: Ashgate.

Patterson, K., Grenny, J., McMillan, R. and Switzler, A. (2002). *Crucial Conversations: Tools for Talking When the Stakes are High*. Hightstown: McGraw Hill.

Paul, J. (1996). Between-method triangulation in organizational diagnosis. *The International Journal of Organizational Analysis*, 4(2), 135–153.

Pidgeon, N. (1998). Safety culture: key theoretical issues. *Work and Stress*, 12(3), 202–216.

Pierluissi, E., Fischer, M., Campbell, A. and Landefeld, S. (2003). Discussion of medical error in morbidity and mortality conferences. *Journal of the American Medical Association*, 290(21), 2838–2842.

Pronovost, P. and Sexton, B. (2005). Assessing safety culture: guidelines and recommendations. *Quality and Safety in Healthcare*, 14, 231–233.

Rochon, T.R. (1998). *Culture Moves: Ideas, Activism, and Changing Values*. Princeton: Princeton University Press.

Rosenfeld, J. (2005). Using the morbidity and mortality conference to teach and assess the ACGME general competencies. *Current Surgery*, 62(6), 664–669.

Rundmo, T. and Hale, A. (2003). Managers' attitudes towards safety and accident prevention. *Safety Science*, 41, 557–574.

Schein, E.H. (1993). On dialogue, culture, and organizational learning. *Reflections*, 4(4), 27–38.

Schein, E.H. (1999). *Process Consultation Revisited: Building the Helping Relationship*. Reading: Addison-Wesley.

Schwarz, R. (2002). *The Skilled Facilitator* (revised edition) San Francisco: Jossey-Bass.

Scott, T., Mannion, R., Davies, H. and Marshall, M. (2003). The quantitative measurement of organizational culture in healthcare: a review of the available instruments. *Health Services Research*, 38(3), 923–945.

Senge, P.M. (1990). *The Fifth Discipline: The Art and Practice of the Learning Organization*. New York: Doubleday.

Senge, P.M. (2006). *The Fifth Discipline: The Art and Practice of the Learning Organization*. New York: Currency Doubleday.

Senge, P.M., Kleiner, A., Roberts, C., Ross, R.B. and Smith, B.J. (1994). *The Fifth Discipline Fieldbook: Strategies and Tools for Building a Learning Organization*. New York: Currency Doubleday.

Sexton, J.B. and Helmreich, R. (2004). Frontline assessments of healthcare culture: safety attitudes questionnaire norms and psychometric properties. Austin: The University of Texas Center of Excellence for Patient Safety Research and Practice. Technical Report No. 04–01. Sponsored by the Agency for Healthcare Research and Quality.

Sexton, J.B., Helmreich, R.J., Neilands, T.B., Rowan, K., Vella, K., Boyden, J., Roberts, P.R. and Thomas, E.J. (2006). The safety attitudes questionnaire: psychometric properties, benchmarking data, and emerging research. *BMC Health Services Research*, 6, 44.

Shackford, J. (2005). Changing airline culture: addressing the behavioral side of change. Presented at the 2005 Symposium on Managing Safety, Reliability, and Services, April 25–27, Incheon, Korea. Retrieved from <www.atwonline.com/resources/whitePapers/track.cfm?wpID=2> on January 15, 2010.

Singla, A., Kitch, B., Weissman, J. and Campbell, E. (2006). Assessing patient safety culture: a review and synthesis of the measurement tools. *Journal of Patient Safety*, 2(3), 105–115.

Sloat, K. (1996). Why safety programs fail ... and what to do about it. *Occupational Hazards*, 58(5), 65–70.

Smits, M., Christiaans-Dingelhoff, I., Wagner, C., van der Wal, G. and Groenewegen, P. (2008). The psychometric properties of the hospital survey on patient safety culture in Dutch hospitals. *BMC Health Services Research*, 8, 230.

Sorra, J. and Nieva, V. (2002). Psychometric analysis of MERS-TM hospital transfusion service safety culture survey. Rockville: Westat, under contract to Barents/KPMG, Contract No. 290-96-004. Sponsored by the Agency for Healthcare Research and Quality.

Sorra, J. and Nieva, V. (2003). Psychometric analysis of the hospital survey on patient safety culture. Rockville: Westat, under contract to Bearing Point, Contract No. 290-96-004. Sponsored by the Agency for Healthcare Research and Quality.

Sorra, J. and Nieva, V. (2004). Hospital survey on patient safety culture. Rockville, MD: Westat, under Contract No. 290-96-0004. Sponsored by the Agency for Healthcare Research and Quality.

Spradley, J. (1979). *The Ethnographic Interview*. New York: Wadsworth.

Stewart, D., Shamdasani, P. and Rook, D. (2006). *Focus Groups: Theory and Practice*. Thousand Oaks: Sage.

Stubbs, I.R. (1998). A leverage force: reflections on the impact of servant-leadership. In: L. Spears (ed.), *Insights on Leadership: Service, Stewardship, Spirit and Servant-leadership* (pp. 314–321). New York: John Wiley.

Taylor, J.C. and Christensen, T.D. (1998). *Airline Maintenance Resource Management: Improving Communication*. Warrendale: SAE Press.

Taylor, J.C. and Patankar, M.S. (1999). Cultural factors contributing to the successful implementation of macro human factors principles in aviation maintenance. In: R. Jensen (ed.), *Proceedings of the Tenth International Symposium on Aviation Psychology*. Columbus: Ohio State University.

Thomas, E., Sexton, J. and Helmreich, R. (2004). Translating teamwork behaviors from aviation to healthcare: development of behavioral markers for neonatal resuscitation. *Quality and Safety in Healthcare*, 13, 57–64.

Triola, N. (2006). Dialogue and discourse: are we having the right conversations? *Critical Care Nurse*, 26, 60–66.

van den Berg, P. and Wilderom, C. (2004). Defining, measuring, and comparing organizational cultures. *Applied Psychology: An International Review*, 53(4), 570–582.

Vogus, T. and Sutcliffe, K. (2007). The safety organizing scale: development and validation of a behavioral measure of safety culture in hospital nursing units. *Medical Care*, 45(1), 46–54.

Weick, K. (1979). *The Social Psychology of Organization*. New York: Random House.

Wiegmann, D., Zhang, H., von Thaden, T., Sharma, G. and Mitchell, A. (2002). Safety culture: a review (Technical Report ARL-02-3/FAA-02-2). Atlantic City: Federal Aviation Administration.

Zremski, J. (2010). Colgan is again tightening fatigue rules – says more-lenient policy after crash is being abused. *The Buffalo News: City and Region*, January 6. Retrieved from <http://www.buffalonews.com/cityregion/story/910797.html>.

Chapter 3
Safety Performance

Introduction

Safety Performance refers to the tip of the Safety Culture Pyramid. Events such as accidents, incidents, and errors, as well as the individual human behaviours that may be safe or unsafe practices, are all part of the Safety Performance layer of the Safety Culture Pyramid. One popular way of thinking about accidents, incidents, and errors is the Heinrich Ratio or the Iceberg Model. Figure 3.1 illustrates the relationship between the Safety Culture Pyramid and the Heinrich Ratio. According to the Safety Culture Pyramid, safety performance refers to observable behaviours that may result in errors that are stopped before they result in any harm or may manifest themselves in undesirable events such as incidents or accidents. The original Heinrich Ratio (Heinrich, 1931), relating one major injury to about 29 minor injuries and about 300 non-injury events, has been popular for decades, and occasionally researchers have tried to test the ratio, but now Taxis, Gallivan, Barber, and Franklin (2006) provide a mathematical argument that the 300:29:1 ratio is a static ratio and does not make much sense in terms of changing the probability of either fatal accidents or incidents, if one were to try to influence the errors at the base of this ratio.

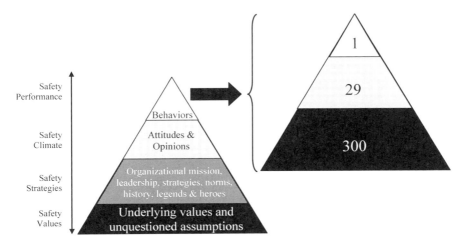

Figure 3.1   The Safety Culture Pyramid and the Heinrich ratio

Reason (1997) used the 'Swiss Cheese Model' of accident causation to describe how a particular error trajectory may pierce through multiple defenses and ultimately result in an accident. Since accidents are relatively infrequent events, and may even appear to be random, it is generally believed that the defense mechanisms are usually effective in blocking the error trajectory from manifesting itself in the form of a fatal/catastrophic accident. One could consider this perspective as a 'vertical' view of the Heinrich Ratio, wherein the majority of the error trajectories are blocked within the bottom layer and therefore do not result in any undesirable outcomes; some escape to the mid layer and result in incidents or minor damage; and very few manage to pierce through all the defense mechanisms and result in a fatal/catastrophic accident. This perspective of the Heinrich Ratio is shared by the Threat and Error Recognition Model developed by Helmreich et al. (1999). Consequently, there is a fundamental assumption that causal factors associated with undesirable events at the middle and bottom tier of the Heinrich Ratio are often similar to, or the same as, the causal factors associated with the top tier events (fatal accidents). Therefore, safety professionals have focused their attention on analyzing the causal factors associated with all undesirable events and developing appropriate safety risk management strategies. Also, since errors and incidents are relatively frequent, they can be tracked and effectiveness of mitigation strategies can be ascertained regularly.

In this chapter, we start with a discussion of Morbidity and Mortality (M&M) Conferences because they are triggered by an undesirable event or they offer a learning opportunity. M&M Conferences and event investigations or debriefing mechanisms in other industries may choose to use Root Cause Analysis (RCA) as a method to conduct retrospective analysis of accidents and incidents. Typically, RCA is triggered by an undesirable event; it tends to identify several systemic factors that may have contributed to the event, and it results in modifications to safety strategies, organizational structures, policies and procedures, or training. The outcomes of an RCA can be indicative of the current state of safety culture at a given organization – secretive culture, blame culture, reporting culture, or just culture. Next, we present two examples to illustrate how RCA of accidents/incidents could be linked with relatively routine errors that do not result in serious harm; thereby pushing the discussion to the lower levels of the Heinrich Ratio. Then, we discuss the impact of safety performance from three perspectives – individual (psychological), organizational (financial), and systemic (societal). These three perspectives are used to illustrate the fact that poor safety performance may have a profound emotional impact on individual professionals involved in the error, substantial financial impact on the organization, and long-term societal impact. Finally, we present a safety performance management strategy that encourages managers to link unit-level safety performance goals with the corresponding financial metrics as well as with organization-level and possibly industry-level safety performance goals.

Morbidity and Mortality (M&M) Conferences

Orlander, Barber, and Fincke (2002) trace the origins of the contemporary M&M conference to the development of the modern hospital as a site for clinical practice and education. They identify Dr. Ernest Codman as developing an end result scorecard, which he introduced at Massachusetts General Hospital in the early twentieth century to capture each patient's symptoms, diagnosis, treatment, complications, and outcome. He suggested that undesirable outcomes be analyzed for causes, discussed, and corrective action taken. While many physicians resisted, Dr. Codman's ideas influenced the American College of Surgeons, which standardized the scorecard practice in hospitals in 1916. From its inception, the processes, procedures and effectiveness of M&M conferences have received little systematic examination with the exception of some attention from surgery and anesthesia. Kravet, Howell, and Wright (2006) point out that all medical schools with residency training programmes were mandated by the Accreditation Council for Graduate Medical Education (ACGME) to institute M&M conferences on a regular basis beginning in 1983 (e.g. American Medical Association, 1995). Kravet et al. state that the 'primary goal of these sessions is to revisit errors to gain insight without blame or derision' (p. 1192).

Orlander et al. (2002) note limitations surrounding the typical conduct of M&M conferences, which constrain their effectiveness. These include important cases not being submitted for M&M discussion, key medical staff not being present during discussions, and cases being selected for educational interest that did not involve morbidity or mortality. Based on a review of 295 surveys they collected from affiliates of the Association of Program Directors of Internal Medicine, Orlander et al. concluded that that the M&M conference 'lacks explicit goals, methods, or formats and it is not described in the internal medicine literature as having an important educational or institutional role in addressing medical error … there is no general consensus about how the conference should be structured in order to ensure that the discussion of mistakes is fruitful' (p. 1003). If M&M conferences are poorly conducted they can result in humiliation, avoidance, denial, and concealment, which all impede learning from mistakes.

Pierluissi, Fischer, Campbell, and Landefeld (2003) present one of the few observational studies of M&M conferences. In their study trained physician observers assessed 66 internal medicine and 85 surgery M&M conferences. The researchers' findings revealed significant cultural differences between internal medicine and surgery. Compared to surgery, internal medicine conferences contained significantly fewer case presentations with adverse events, significantly more time devoted to invited speakers, and significantly less time used for audience discussion. The researchers also found that both surgery and internal medicine conference leaders seldom made use of explicit language to indicate an error was being discussed and seldom acknowledged having committed an error. They conclude that 'teachers in both surgery and internal medicine missed opportunities to model recognition of error and to use explicit language in error discussion by acknowledging their personal experiences with error' (p. 2838).

Orlander et al. (2002) offer several suggestions for improving M&M conferences, including the following: a clear definition of the conference's purpose (e.g. identify, reflect, and disseminate insights from error to improve medical practice); guiding principles (e.g. goal is learning not criticism); case selection (e.g. choose recent cases); moderator (e.g. senior physician skilled at creating a supportive atmosphere and managing difficult discussions); attendance (e.g. members at all levels of the organizational hierarchy); and conclusions (e.g. clear summary should be given with corrective action steps for improvement).

In an attempt to improve the effectiveness of M&M conferences, Kravet et al. (2006) integrated the M&M conferences into the grand rounds schedule and used the six core competencies from the ACGME's Outcomes Assessment Project (Swing, 1998) to frame the discussion of cases. These six competencies are patient care, medical knowledge, practice based learning improvement, interpersonal communication skills, professionalism, and systems-based practice. Kravet et al. maintain that this approach focuses attention beyond the individual to promote reflection on the role of communication and system failures as well as recognizing professionalism in the handling of adverse events. The authors report that this format is well received and has resulted in the recognition of system deficiencies and subsequent process redesign.

Aboumatar, Blackledge, Dickson, Heitmiller, Freischlag, and Pronovost (2007) studied the conformity of morbidity and mortality conferences to models of medical incident analysis with a survey/interview of 12 clinical department chairs at Johns Hopkins Hospital. While most of the respondents shared a common perception of M&M conference goals (e.g. to review medical management and improve patient safety and quality) they found wide variation in M&M processes with few processes matching established models of medical incident analysis. Aboumatar et al. conclude that an evidenced-based best practice model is needed to improve M&M conferences. They suggest an eight step conceptual M&M conference model that includes case identification, selection, review, analysis, discussion/presentation, summary, recommendations, follow-up, and feedback.

To better address system-wide problems, Deis et al. (2008) implemented a monthly interdisciplinary Morbidity, Mortality and Improvement Conference. A distinctive goal of this approach is a commitment to systematic process change. M&M cases are carefully selected and prepared to include a focus on contributing systems issues with case details summarized in a time series illustration. The conference follows a standardized agenda that analyzes the case by examining six broad areas of influence (i.e. procedure, environment, equipment, people, policy, other). Results from the case discussion are captured in a 'fishbone' diagram to visualize cause and effect relationships for the adverse event. Once agreement is reached on the key causes, action plans are developed and individuals are assigned to work groups to take corrective action within specified timeframes. Progress and results from these work groups are presented at subsequent conferences. As an example of this process Deis et al. found that communication problems contributed to about two thirds of adverse events. To correct this situation an SBAR (i.e. Situation, Background, Assessment and Recommendation) communication method (e.g. Haig, Sutton, and Whittington, 2006; Leonard, Graham, and Bonacum, 2004) was implemented hospital-wide. SBAR provided a common model to structure communication and improved information exchange for better coordination and decision-making among members of the healthcare team.

Berenholtz, Hartsell, and Pronovost (2009) reported the use of a simplified RCA tool to learn from defects, which was employed by fellows from anaesthesiology, surgery and emergency medicine at Johns Hopkins University. The tool provided a structured approach to establish causation, implement corrective action and evaluate the effectiveness of the intervention. The authors concluded, based on 13 applications, that the tool was effective in teaching an analysis of errors (especially system failures), learning from the incidents, improving patient care, and providing a strong foundation for use in the fellow's future practice. Berenholtz et al. believe that this tool offers an important additional approach to learning from errors in traditional M&M conferences.

Szekendi, Barnard, Creamer, and Noskin (2010) report on work at Northwestern Memorial Hospital in Chicago where M&M conferences were used to create a forum for transparent, non-punitive, interdisciplinary dialogue about patient safety problems. These M&M conferences focused discussions on RCA of the event and systems thinking to identify effective solutions. Szekendi et al. (2010) note the importance of reducing the widespread fear of blame and reprisal related to reporting errors that exists in healthcare. They point to findings from the Hospital Survey on Patient Safety Culture (<http://www.ahrq.gov/qual/patientsafetyculture/hospsurvindex.htm>) which show that for all the areas measured, the lowest average favourable response is for the domain of nonpunitive response to error (mean of 43–44 per cent), which has been stable for the preceding three years (2007–2009). They argue that 'employees remain silent if the identification of problems or the suggestion of solutions is seen as either risky or useless' (p. 7). The experience at Northwestern Memorial Hospital shows that effectively structured M&M conferences demonstrate leadership commitment to patient safety and just culture while providing a powerful opportunity to learn from medical error and improve patient safety.

Root Cause Analysis (RCA)

Root Cause Analysis (RCA) is a retrospective analysis of the accident case to uncover the cause and the contributing factors (Wald and Shojania, 2001). While RCA provides a reasonable structure to conduct the analysis, it is not perfect. Wald and Shojania argue that RCA tends to suffer from hindsight bias as well as existing political pressures, as evident from the shift in accident investigations from purely technical failure analysis to a more organizational and human factors oriented analysis. Nonetheless, it is able to identify the cause (however it is very difficult to verify) and the contributing factors. Reason's (1997) model may be used to list the causal factor as the active failure, and list the contributing factors as latent failures. Active failures tend to be human errors of omission or commission or technical failures. Latent failures, on the other hand, tend to be systemic failures that simply stay in a benign state until the right combination of error producing conditions line up. RCA and Reason's model are widely used in both aviation and healthcare.

Contemporary RCA in aviation and healthcare seeks to understand the complex interaction of variables that impact safety performance by investigating the entire system of dynamic linkages between safety behaviour, climate, strategies, and underlying values. Originating decades ago in accident analysis and quality control, RCA is employed today by all high-risk industries. Rather than being a monolithic methodology, RCA employs a variety of techniques and approaches. Wilson, Dell, and Anderson (1993) offer a useful overview of RCA techniques and include discussions of change analysis, barrier analysis, events and casual factors analysis, tree diagrams, management oversight and risk tree analysis, human performance evaluation, and cause and effect fishbone diagrams. Latino and Latino (2006) and Paradies and Unger (2000) also provide helpful advice and directions for carrying out an RCA.

Carroll, Rudolph, and Hatakenaka (2002) discuss how RCA conducted in other high-consequence industries provides important lessons for healthcare. They discuss how the analysis of an explosion at a chemical refinery led to cultural change and continual improvement of safety by challenging behaviours, attitudes, thinking, strategies, and values related to safety. These authors concluded that 'there are strengths of the healthcare culture that can be built upon to support new understandings and new practices, rather than trying to break the existing culture' (p. 268).

RCA has long been used in aviation and aerospace. Several examples in addition to the well-known space shuttle disaster investigations include the following: aerospace accidents (Weiss, Leveson, Lundqvist, Farid, and Stringfellow, 2001); general aviation controlled flights into terrain (Shappel and Wiegmann, 2001); and rule violations by aviation maintenance technicians (Patankar, 2002). Vincent, Taylor-Adams, and Stanhope (1998) argue that broader, deeper, and more systematic investigational approaches to risk and accidents, initially derived from complex industrial and aviation systems, should be applied to healthcare to improve safety. Wilf-Miron, Lewenhoff, Benyamini, and Aviram (2003) discuss how concepts used in aviation to understand and mitigate error, including RCA, could be successfully applied to improve safety in a large ambulatory care organization.

During the past decade, RCA has been more consistently applied to healthcare. Rex, Turnbull, Allen, Vande Voorde, and Luther (2000) applied RCA to investigate adverse drug events in a tertiary referral hospital. They found that process changes implemented as a result of the RCA were related to a 45 per cent decrease in voluntarily reported adverse drug events. The Department of Veterans Affairs has made the most extensive institutionalized use of RCA. Bagian, Gosbee, Lee, Williams, McKnight, and Mannos (2002) reported that by August 2000 a total of 173 Veterans Affairs facilities had implemented RCA of adverse events and 'close calls'. These authors describe their RCA and corrective action methodology and its real-world application in two case studies: one involving hazards in a magnetic resonance imaging room and another case with a cardiac pacemaker malfunction. Using RCA, their investigations focused on identifying actionable system vulnerabilities. Vincent (2003) also offers suggestions for investigating medical incidents, capturing critical lessons learned, and supporting the patients, families and staff members involved. He points out that typically there are chains of causes that lead up to the critical event and therefore he strongly recommends developing a systems-level analysis that is proactive and forward looking. Woolf, Kuzel, Dovey, and Phillips (2004) also focus on understanding strings of mistakes and describe a cascade analysis technique for capturing these relationships and preventing future medical errors.

Battles, Dixon, Borotkanics, Rabin-Fastmen, and Kaplan (2006) offer a sensemaking framework to learn from safety related events. They argue that sensemaking serves as a useful conceptual framework to integrate analyses from the single event level with the process level and system level. They maintain that sensemaking conversations about the event increase ownership and effective action to mitigate future risk. Wu, Lipshutz, and Pronovost (2008) discuss the effectiveness and efficiency of RCA in healthcare. They note that RCA is widely used but may be performed incorrectly or incompletely. Lessons learned may not be integrated across investigations. They conclude that 'formulating corrective actions is more difficult than finding problems, and follow-up on outcomes is rare' (p. 686). They recommend that RCA should be carefully evaluated for effectiveness and utility. They suggest that a special emphasis should be placed on assessing the impact of changes implemented to improve safety so that best practices can be established and disseminated.

Overall, RCA is a rigorous process of extensive data collection (mostly qualitative data), recreation of the entire timeline of events and actions, reconstruction of the debris for technical evaluation, and background investigation of the personnel (including qualifications, training, and health conditions), patient records or aircraft maintenance records, equipment/technology issues, advisory documents/alerts, regulatory requirements, etc. As the details about the accident are reconstructed, the RCA process calls for description of the chain of events that ultimately led to the accident. In this process, active failures are identified. Then, the rationale for those errors, as well as any safety nets that might have been in place, is investigated. The findings of this part of the investigation typically uncover latent failures in the system. The causal pathway continues to drill deeper until none of the final reasons for failure is a human error. The five rules of causation[1] are as follows:

---

1 <http://www.nyhealth.gov/nysdoh/healthinfo/patientsafety.htm>.

1. Root cause statements must clearly show the 'cause and effect' relationship.
2. Negative descriptions should not be used in root cause statements.
3. Each human error must have a preceding cause.
4. Violations of procedure are not root causes – they must have a preceding cause.
5. Failure to act is only causal when there is a pre-existing duty to act.

Example 3.1: Comair Flight 5191

On August 27, 2006, Comair Flight 5191 crashed during takeoff from Blue Grass Airport, Lexington, Kentucky. According to the NTSB's factual report,[2] the aircraft lined up on the wrong runway. Instead of taking off from Runway 22, it took-off from Runway 26, a much shorter, unlit, and non-operational runway. Figure 3.2 illustrates the intended (dotted line with arrow) versus actual (solid line with arrow) path of the aircraft.

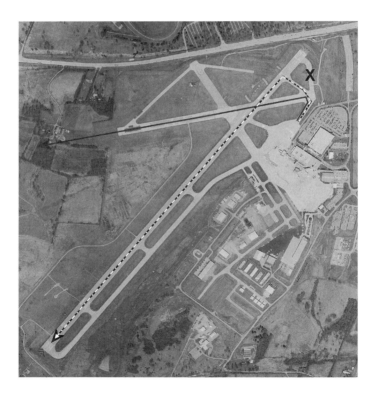

Figure 3.2   Runway configuration at Blue Grass Kentucky Airport

2   <http://www.ntsb.gov/ntsb/major.asp; NTSB ID DCA06MA064>.

The NTSB established causal relationship between the act of lining up with the wrong runway (cause) and the accident (effect). Next, we must drill deeper and try to determine why the crew lined up with the wrong runway and also why they may not have been able to stop or correct themselves. The factual report identifies several factors that may have contributed to the crew lining up with the wrong runway:

- The crew was engaged in non-essential conversation during the crucial moments of taxiing; consequently, they may have been sufficiently distracted.
- The first officer, who was in charge of the take-off, was noted to be yawning and missing certain checklist items.
- The captain was in charge of the taxi operation and held short of Runway 26 instead of continuing on through and holding short of Runway 22. As you can see from Figure 3.2, the crew had to cross Runway 26 to get to Runway 22. The NTSB report claims that the crew believed that they were lined-up for Runway 22.
- The Air Traffic Controller should have monitored the flight until he handed it off to the next controller. Failing to monitor the flight all the way through take-off, the controller was unable to stop the aircraft from taking off from the wrong runway.

As is customary with such accident reports, the analysis is very focused on the specific event and the factors that could be specifically linked with the final outcome (accident). In order to conduct a sufficiently detailed review, the following key aspects are considered in the full report: equipment integrity, maintenance history, crew personal health history, crew training and specific errors associated with the flight, air traffic controller role, weather conditions, and organizational factors.

In terms of the Heinrich Ratio, the Comair accident clearly represents 'fatal accident' at the top of the triangle. Underneath, however, is the layer of incidents, which asks the question of how often pilots line up on the wrong runway, take off from the wrong runway, or land on the wrong runway. All such incidents are considered runway incursions. Technically, a runway incursion is an incorrect/unauthorized presence of an aircraft, vehicle or person on a runway or helicopter landing pad. In the Comair accident, since the aircraft took off from the wrong runway, it is classified as a runway incursion. A review of the runway incursion trend reveals that nationally, there are about 200–300 runway incursions every quarter; there was a slight decrease in the total events in FY2009 compared to FY2008; and the FY2008 and FY2009 trend lines are almost identical. Figure 3.3 presents the latest runway incursion trend by quarters.[3]

The fact that Comair 5191 resulted in a fatal accident is egregious, but on a national scale, there are about 1,000 runway incursions a year that do not result

---

3   <http://www.faa.gov/airports/runway_safety/statistics/year/?fy1=2009andfy2=2008>.

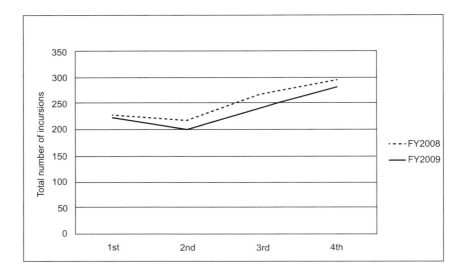

Figure 3.3 **Runway incursion trend by quarter**

in fatal accidents. The FAA's Runway Safety Program[4] is dedicated to reducing runway incursions. Special awareness training for pilots, controllers, and ground vehicle operators, improvement of airport signage, and ongoing research are part of this programme. In the United States there are about 10,000,000 air carrier departures per year.[5] Even if we don't consider the additional general aviation traffic, we are looking at about one runway incursion in every 10,000 air carrier departures.

The active failures in this case were the facts that the flight crew did not catch the error in time to stop the aircraft. Also, there was an active error on the part of the controller because he did not monitor the aircraft all the way through takeoff. Neither the pilots nor the controller repeated the runway number when they were last cleared for takeoff. Of course, there were several latent failures in the system: the runway layout and the taxiway path was confusing and prone to errors, the pilots were tired and possibly in a hurry to get home, and there were no gates or other hard markings on the closed runway to minimize the chance of a wrong runway takeoff.

Going back to the Heinrich Ratio, it is widely acknowledged that in order to minimize such accidents, we need to focus on reducing the runway incursion incidents. Therefore, improvement of runway safety is on NTSB's 'Most Wanted' list. The FAA monitors runway incursions on a daily basis and also has an Office

---

4   <http://www.faa.gov/airports/runway_safety/>.
5   <http://www.bts.gov/publications/national_transportation_statistics/html/table_01_34.html>.

of Runway Safety to specialize in providing dedicated services to improve runway safety. Besides the FAA and the NTSB, the airlines, the pilots unions, the Air Transport Association, Air Traffic Control, and airports are all working together to minimize runway incursion.

Example 3.2: Heparin Administration Error

The Agency for Healthcare Research and Quality (AHRQ) has developed an online database of Morbidity and Mortality. All the reports are factual, but anonymous. The latest reports and expert commentary can be found at <http://www.webmm.ahrq.gov/>. The case presented here is available on the website as Case ID 201:

> A man was admitted to the hospital for an elective left carotid endarterectomy. This is a surgical procedure to remove plaque from the carotid artery to reduce blockage and thereby reduce chances of a stroke. The surgeon typically clamps the artery on both sides of the suspected area of blockage, makes an incision in the artery, removes the plaque, and sutures the artery shut. During this procedure, it is critical that the blood does not clot; hence an anticoagulant is injected. In this case, the anesthesiologist mistakenly administered 120 units of heparin instead of 12,000 units because he used the wrong vial – he mistakenly used 10 units/mL vial rather than 1,000 units/mL. There was no obvious clinical damage in this case; however, the error was serious.

According to Dr. Vanderveen,[6] one problem is that the two vials of heparin are colour differentiated with different shades of the same colour. The other problem is that the nurses tend to pick between the two types of vials based on their physical location in the Automated Dispensing Cabinet (ADC) while the anaesthesiologists differentiate them based on the two shades of the base colour. Ironically, the purpose of the ADC is to prevent administration of incorrect medication.

Applying RCA and Reason's Model, it is clear that the causal factor in this case was accidental administration of a low concentration medication – it was a medication error or an adverse drug event at the hands of the anaesthesiologist. But the RCA approach would ask the question why? Why did the anaesthesiologist pick up the wrong vial, why did he not recognize the error, could anyone else have stopped him? Well, we know that the two vials tend to look very similar – that's a latent failure that requires the administrators to be on high alert. If we consider this accident to be at the top of the Heinrich ratio, then we should look at how often nurses and anaesthesiologists administer a wrong dose of heparin. Further, if the causal factor can be generalized to an adverse drug event, then how often do we have those? The error pathways for most adverse drug events are one of the following: confusing packaging, dilution errors, contamination, similar sounding names, or stocking error, and failure of defenses (requirement for staff to perform

---

6 <http://www.webmm.ahrq.gov/case.aspx?caseID=201>.

the 'five rights'[7] – right patient, the right drug, the right dose, the right route, and the right time).

The above case (Case ID 201) was posted in May 2009. However, the issue of incorrect administration of heparin is not new. Some of the relatively recent cases related to heparin dose errors are as follows:

- *September, 2006*: three infants received overdose of heparin at Methodist Hospital in Indianapolis; two of them died at Methodist and one died at Riley Hospital.[8] Baxter Healthcare Corporation issued a medical safety alert to remind medical practitioners not to rely on two different shades of blue labels when picking the drug – read the concentration of the drug prior to administering.
- *November, 2007*: accidental administration of adult heparin to actor Dennis Quaid's twins nearly killed the two babies.[9] Instead of receiving 10 units of heparin, the babies got 10,000 units.
- *July, 2008*: seventeen premature babies received an overdose of Heparin (paediatric; not like the above two cases) at CHRISTUS Spohn Hospital in Corpus Christi, Texas. The error occurred in the mixing process in the hospital's pharmacy.

The Joint Commission's Sentinel Event Database includes, '446 medication-related sentinel events (9.3 per cent of all events) reported from January 1997 through December 1997, with 7.2 per cent of these involving anticoagulants; of those, two-thirds (21) involve heparin' (Joint Commission, 2008). Therefore, 'anticoagulants have been identified as one of the top five drug types associated with patient safety incidents in the United States'. According to a special report prepared by US Pharmacopeia[10] for the *Los Angeles Times*, 'between 2001 and 2006, more than 16,000 heparin errors were blamed on incorrect dosing' (Ornstein, 2007). The Joint Commission reported that 'between 1997 and 2007, heparin was the most common drug involved in anticoagulant administration errors and out of 34 patients involved, 28 died and six suffered from loss of function'.

Clearly, heparin errors have been common and have resulted in some serious consequences. A review of the heparin errors presented here reveals that there are two kinds of heparin errors: the first is a contamination issue that is either at the manufacturer level or at the pharmacy level; the second is the administration error. The contamination issue has been addressed by increasing quality control measures by the manufacturer/pharmacy. The administration issue has been addressed by improved packaging and intensified awareness campaigns. Now, the

---

7 <http://www.ihi.org/IHI/Topics/PatientSafety/MedicationSystems/ImprovementStories/FiveRightsofMedicationAdministration.htm>.
8 <http://www.theindychannel.com/news/9869794/detail.html>.
9 <http://abcnews.go.com/GMA/story?id=3956580andpage=1>.
10 <http://www.usp.org/hqi/patientSafety/medmarx/>.

10,000 USP vial comes with additional packaging to positively differentiate it from the lower strength doses.

Impact of Safety Performance

Psychological Impact on the Individual

The reporting of medical errors has traditionally focused on the cognitive dimensions of conducting a causal analysis of the adverse event and action planning to prevent similar mistakes in the future. An emerging area of study now investigates the significant emotional cost of errors for physicians and the related impact on patient care.

Authors such as Jaques (1955) and Menzies (1960), working from a psychoanalytic perspective, developed and extended the concept of social systems functioning as psychological defence mechanisms to deal with anxiety. They discussed topics such as nurses' detachment and denial of feelings and attempts to eliminate decisions by ritual task performance. Bain (1998) discussed the importance and difficulty of altering such maladaptive social defences to overcoming resistance to change and to promoting learning from organizational experience.

Mizrahi (1984) identified denial, discounting (e.g. blaming others, superiors, or the system), and distancing (e.g. appeals to human limitations) as the three major mechanisms that internists-in-training used to deal with the psychological impact of self-doubt and guilt related to medical error. Mizrahi argues that these physicians see themselves 'as the sole arbiter of mistakes and their adjudication' (p. 146). This belief provides psychological defence, guards the physicians' autonomy, depends on complex technical expertise, and is embedded in a larger system that is reluctant to judge a physicians' actions unless there is litigation or criminal negligence.

Christensen, Levinson, and Dunn (1992) conducted individual, in-depth interviews with eleven physicians who were asked to discuss a previous medical error that they had committed, their beliefs about the cause of the mistakes, the emotions they experienced, the coping techniques they used, and resulting changes in their practice of medicine due to the errors. These researchers found errors to be common, yet the mistakes were seldom disclosed to colleagues. The emotional impact on the physician was great, yet little support was found from medical peers. This research also revealed that even after several years, physicians often recalled their mistakes in vivid detail.

Christensen et al. (1992) found that when medical mistakes occurred physicians reported intense emotions of shock, agony, and anguish. Physicians' 'feelings of fear, guilt, anger, embarrassment, and humiliation were unresolved ... even a year after the mistake' (p. 426). Physicians faced important decisions about disclosing their mistakes. When the error was shared with colleagues the focus was typically on problem solving to prevent a reoccurrence of the mistake and not on the emotional experience. If physicians' disclosed their emotions related to the error it was most likely to be shared with a spouse or close friend. Physicians often felt emotionally isolated by their mistake. Christensen et al. conclude that the 'emotions of shame, guilt, depression, and anxiety may be influenced both by perfectionism ... and by the way doctors are socialized into the profession' (p. 430). While the physicians reported that they had learned from their mistakes and improved their practice, the researchers argue that better processes should be in place not only to learn from medical error, but also to support the emotional recovery of the physicians.

Newman (1996) studied the emotional adversity of medical errors on family physicians by analyzing how 30 doctors recounted their most memorable mistakes. The most common emotions identified were self-doubt, disappointment, self-blame, shame, and fear. The most valued form of support was being able to speak to another person about the adverse event; however, about two thirds of the physicians received this support from individuals who were not their colleagues. Newman states, 'most of the physicians were unable to talk to their peers and only conditional willingness to support a colleague suggests that to admit imperfection and share fallibility, however, is socially unacceptable' (p. 74).

Houston and Allt (1997) studied the experience of 30 junior physicians early in their hospital careers who were practicing in the United Kingdom. Statistically significant increases in self-reported anxiety, insomnia, and somatic symptoms were found. Additionally, significant increases were found for these junior doctors in the number of self-reported medical errors and errors made in their everyday life outside of the medical context. The researchers suggest that working conditions and hours of service for the junior doctors should be carefully examined to reduce the occurrence of psychological distress and errors while improving the transition to increased professional medical responsibilities.

In a prospective longitudinal study, West et al. (2006) assessed the relationship between resident physicians' self-perceived medical errors and measures of quality of life, burnout, depression, and empathy. Responses were collected from 184 residents who completed surveys on a quarterly basis. Findings revealed that 34 per cent of the residents reported at least one major medical error. Self-perceived medical errors had significant adverse correlations with quality of life, all domains of burnout (i.e. depersonalization, emotional exhaustion, and lack of personal accomplishment), and empathy. Self-perceived errors were significantly related to an increased likelihood of depression. Furthermore, increased burnout and reduced empathy were associated with an increased likelihood of self-perceived errors in a subsequent quarterly reporting period.

West et al. (2006) conclude that perceived errors and negative emotions may be related in a vicious reciprocal cycle of events, 'whereby medical errors may lead to personal distress, which then contributes to further deficits in patient care' (p. 1075). However, it was also found that increases in measures of personal accomplishment and empathy were related to a decreased likelihood of self-perceived errors in a subsequent reporting period. West et al. suggest addressing system-wide issues to remove organizational-structural sources of error.

Nonetheless, when errors do occur it appears that physicians have inadequate resources for support and 'strategies for coping with the emotional impact of errors are needed but have been slow to develop' (p. 1076). Delbanco and Bell (2007) discuss the importance of developing an organizational structure to restore communication and provide emotional support. They suggest 'first steps might include creating structured curricula for professionals addressing both error prevention and response, removing stigma from transparent reporting systems, and deploying a system of expert first responders who guide patients and clinicians when an error occurs' (p. 1683). Likewise, to improve physicians' responses to medical error, Engel, Rosenthal, and Sutcliffe (2006) suggest that opportunities should be seized to discuss occasions when even though errors occurred, nonetheless, the patient still had a favourable outcome.

Gallagher, Waterman, Ebers, Fraser, and Levinson (2003) used focus groups and content analysis to study patients' and physicians' attitudes toward the disclosure of medical errors. They found that patients and physicians had strikingly different opinions about the information that should be shared about medical errors. While patients wanted full disclosure of information about the error, physicians were likely to choose their words very carefully due to fear of litigation and concern about damage to their professional reputation. Gallager et al. also found that patients desired an apology from physicians; however, physicians were concerned about the legal liability created by an apology.

Lazare (2006) discusses the practice of genuine apology in healthcare as an emerging clinical skill that can produce significant healing. He argues that for the physician an apology can reduce the negative emotions of guilt, shame, and fear of retaliation; while for the patient, receiving an apology may diminish anger and lead to forgiveness and reconciliation. Lazare describes a number of psychological processes through which an apology can produce healing, including the following: restoring self-respect and dignity, demonstrating concern, restoring a balance of power in the patient-physician relationship, and the demonstration of contrition and suffering. Beyond the ethical, professional, and humanitarian goals of apology, it has frequently been the case that when sincere apologies are given there is often a corresponding reduction in the number and financial costs of malpractice lawsuits. However, insincere apologies can fail and can make the situation worse. Physicians may resist admitting error and offering apologies; however, Lazare concludes 'that admissions of harm and apologies strengthen, rather than jeopardize, relationships and diminish punitive responses' (p. 1403).

These studies of emotion related to healthcare error indicate that physicians pay a high psychological price for concealing or minimizing discussion of medical errors. As recognized by Jaques (1955) and Menzies (1960) social structures in complex professional organizations develop to help members handle anxiety. However, this capacity may either be pathological or productive, depending on the maturity of the organization's safety culture. Productive cultural responses to error, and the emotions and anxiety aroused, can be handled in more direct ways through reporting and just cultures than the response provided by secretive and blame cultures. The journey from pathological to productive approaches to error can be facilitated by focusing on appreciative inquiry (e.g. Fryer-Edwards et al. 2007; Shendell-Falik, Feinson, and Mohr, 2007) that begins with discussions of 'good saves' and stories of success. Attempts to change culture that focus exclusively on negative events may exacerbate the anxiety related to error, increase resistance, and shut down a sense of urgency to change the system.

Financial Impact on the Organization

The National Quality Forum[11] has identified 28 Serious Reportable Adverse Events or 'Never Events' – these are medical errors or conditions that are preventable. Some states are starting to refuse payments for costs associated with additional care involving some or all of the never events. Some of these items are as follows:

- Surgery on wrong body part.
- Surgery on wrong patient.
- Wrong surgery on a patient.
- Foreign object left in patient after surgery.
- Post-operative death in normal health patient.
- Death/disability associated with use of contaminated drugs, devices or biologics.
- Infant discharged to wrong person.
- Death/disability associated with medication error.
- Death/disability associated with incompatible blood.
- Maternal death/disability with low risk delivery.
- Stage three or four pressure ulcers after admission.
- Death/disability associated with a fall within facility.

Each of these events has an associated additional cost – both the direct cost of correcting the error, redoing the surgery; and the indirect cost of additional length of stay in the hospital, additional medication, etc. In an effort to encourage hospitals to adopt best practices and eliminate preventable errors, Medicare and some insurance companies are refusing to pay costs associated with preventable medical errors. These costs will have to be absorbed by the hospital. For example, nosocomial infections, or hospital-acquired infections, account for approximately two million infections per year (100,000 deaths per year) in the United States, and they cost an estimated $30.5 billion each year, which is about $15,275 per infection (McCaughey, 2008). These infections are preventable; therefore, the deaths and the medical costs are avoidable.

A case study of Ascension Health's efforts to eliminate nosocomial infections reported seven catheter-related blood stream infections (CR-BSI) per 1,000

---

11 <http://www.qualityforum.org/Publications/2008/10/Serious_Reportable_Events.aspx>.

centreline catheter days and Ventilator Acquired Pneumonia (VAP) was at 8.2 per 1,000 ventilator days (Berriel-Cass, Adkins, Jones, and Fakih, 2006). The average cost to treat CR-BSI is $91,000.[12] That's approximately $637,000 for CR-BSIs and about $125,255 for VAPs. Together, the estimated cost of these infections would be about $762,255 per 1,000 catheter/ventilator days. Through strong commitment from the top leadership, focused training, and intense discipline, the Ascension team was able to bring the CR-BSI rate down to 3.15 per 1,000 days and the VAP 3.3 per 1,000 days. Such significant reduction in infections was particularly timely because beginning October 1, 2008, Medicare and Medicaid have started declining reimbursements for hospital-acquired infections.[12] Therefore, all such costs will have to be absorbed by the hospital. This stance by Medicare/Medicaid is tantamount to consumers refusing to pay for unsafe service/products or switching from a provider of 'allegedly unsafe' service/product to a provider of 'apparently safe' service/product.

The notion of switching from one provider to another is possible to study in the airline sector as well. In this context, one should ask whether consumer behaviours reward safety. In other words, if one provider has an accident, will consumers choose a different provider of the same/similar service (is there a switching effect)? And, if one provider has an accident, do other providers in the same business sector suffer a financial loss (is there a negative spillover)? Bosch, Eckard, and Singal (1998) studied the impact of airline accidents (1978–1996) on stock values and concluded that there was evidence to support both a switching effect and a negative spillover. That is, in the event of an airline accident, non-crash airlines gain revenue (consumers tend to choose 'non-crash' airlines) and consumers tend to fly less (all airlines within that service sector experience loss of revenue). Patankar and Mondello (2010) used a similar approach with a sample of five airlines for accident and stock values between 1995 and 2003. Highlights of their findings are as follows:

- The combined trend for all five airlines indicates a dip in the stock value after every airline accident, except after American Airlines flight 587.
- The stock price for ValuJet (later known as AirTran) did not recover to the pre-flight 592 stock price.
- All comparison airlines, except AirTran, lost stock value immediately after TWA 800 crash (AirTran does/did not have Boeing 747 aircraft).
- Alaska Airlines' stock was on the decline prior to the flight 261 accident, but it did not recover to pre-flight 261 level.
- All airlines lost stock value immediately after September 11, 2001.
- American Airlines continued to gain stock value after flight 587, but within a month, it started losing value and did not recover within the review period.

---

12 <http://www.firstdonoharm.com/HAC/CRBSI/>.

Patankar and Mondello (2010) also discovered that US Airways lost nearly 25 per cent of its value after flight 1549 and the stock price had not recovered to pre-crash rate for at least one year. So, at least in aviation, accidents can have catastrophic and long-term impact on the financial viability of the company. From a purely financial perspective, the customers – whether they are insurance companies or individual consumers – are starting to demand a standard of safety that will be essential for the financial viability of hospitals and airlines.

Societal Impact

In the case of the Comair accident, the NTSB ruled that the pilots' failure to 'recognize that they were initiating a takeoff on the wrong runway' contributed to the accident. Further, the air traffic controller did not monitor the flight all the way through takeoff. According to a 1993 report filed with NASA's Aviation Safety Reporting System, pilots of a commercial air carrier did exactly what the Comair pilots did – they lined up to take off from Runway 26 instead of Runway 22 (Stewart, 2007). The difference, however, was that both the crew as well as the air traffic controller noticed the problem and averted a tragedy.

A number of legal suits were filed in response to the Comair accident. Families of the passengers filed suits against Comair for negligence; whereas Comair filed suits against the Lexington Blue Grass Airport and the Federal Aviation Administration to seek a contributory share in the responsibility for the accident. One of the most startling developments from the Comair accident was that in an effort to strengthen the suit against Comair, the plaintiffs' lawyers sought copies of voluntarily submitted pilot reports.[13] The US Magistrate Judge James B. Todd granted that request. Judge Todd's ruling challenged the aviation industry's sacred achievement – interpersonal trust that has been built between employees, management, and regulators through voluntary reporting of human errors. One of the foundational principles of this voluntary reporting programme is that it provides protection from federal regulatory action and in some cases corporate disciplinary protection. In exchange, the company and the regulator can learn about the systemic issues contributing to the accident/incident and address the real root causes. Ironically, the purpose of the error reporting system is to drive the organizational culture away from blame; now, it could be used to blame the company as well as the FAA. While the use of the system in a completely unintended way is ironic and unfortunate, it poses the question about corporate responsibility – what is the organization or the government's responsibility toward addressing known safety hazards? Fortunately, the error reporting programmes continued to grow and all three parties – employees, management, and the FAA – continued to support the programme.

---

13   <Aviation Safety Action Program Reports: http://www.faa.gov/about/initiatives/asap/>.

In the case of the heparin accidents involving Dennis Quaid's twins, the hospital acknowledged three safety lapses: the pharmacy technician took the heparin without having a second technician verify its concentration, a different technician at the paediatric unit failed to verify the heparin concentration, and finally the nurses who administered the heparin also neglected to verify that it was the correct concentration (Ornstein, 2007).

Again, several legal suits were filed. The Quaids filed suit against Baxter Healthcare Corporation because 'it failed to recall the high-concentration vials and used similar colour backgrounds for labels of both high- and low-concentration vials' (Ornstein, 2007). Ceders-Sinai took disciplinary actions against the pharmacists and nurses involved in the heparin errors. Although several hospitals participate in voluntary reporting of medication errors in the MEDMARX database,[14] which is an international database, there is no legal or corporate protection for employees submitting such information. So, in the absence of such explicit protection, the hospitals and the medical equipment manufacturers or pharmaceuticals have to work together to build an environment of trust and mutual accountability. In contrast, and in response to the death of three infants in September 2006, in Indianapolis, Methodist Hospital issued a statement on September 20, 2006. According to that statement, the employees involved were placed on leave, but they were being assisted and supported; they were expected to return to work. Clarian Health, the parent company of Methodist Hospital, regarded the heparin error as an 'institutional error – one in which our procedures failed. Those procedures now have been, and are continuing to be, reevaluated. A number of measures to safeguard against such an error occurring again have already been put in place'.[15]

In both cases, the post-accident challenges were fraught with legal battles that were partly aimed at assigning blame and partly at improving the system for the future. The airline industry was challenged to maintain its trust in the error reporting systems, regardless of the legal protection, and it did. The healthcare system was challenged to acknowledge the systemic faults and move toward eliminating the avoidable errors, and it is making significant progress in this direction. Overall, society is seeking a new balance point – a point where full disclosure and accountability will co-exist and both flight safety and patient safety will continue to improve.

Safety Performance Targets

Helmreich et al. (1999) use their Threat and Error Recognition Model to argue that threats and errors exist all the time; the successful management of these threats and errors could be the difference between accident and prevention. Similarly, Reason

---

14   <https://www.medmarx.com/>.

15   Heparin error backgrounder drafted September 20, 2006. Available at <http://www.postdoc.medicine.iu.edu/>.

(1997), Dekker (2007), and Marx (2009) argue that the underlying behaviours, regardless of the outcome, need to be addressed because such an approach is consistent with the Just Culture paradigm. For example, a centreline infection may be caused by a nurse not observing adequate hand hygiene; however, that nurse's underlying behaviour of not disinfecting hands prior to handling of the centreline is a violation of protocol, regardless of whether the behaviour results in centreline infection. Further, since centreline infections could lead to extended hospital stays and/or death of the patient, centreline infection is a good safety performance metric to track.

Safety performance management strategy could be classified as reactive, proactive, or predictive. One could consider these categories as progressive improvements or as a measure of maturity on the safety performance management continuum. Integral to this strategy is the data acquisition and analysis capability. In a reactive strategy, the undesirable event serves as the trigger and a RCA approach is used to assemble the relevant data; however, the data tend to be focused on a specific event and on the historical trend of precursors. In a proactive strategy, safety performance data are collected as a matter of standard and routine practice, and systemic issues are addressed prior to the occurrence of an undesirable event. Both qualitative and quantitative data can be integrated and proactively analyzed to identify the key issues. In a predictive strategy, the data analysis is significantly sophisticated; multiple sources and types of data are integrated and advanced data mining tools are used to discover unique patterns of coincidences that are typically difficult to identify. Patankar and Taylor (2004) called such coincidences 'low-frequency – high-consequence' events.

Reactive Safety Performance Management Strategy

A reactive safety performance management strategy is typically triggered by a high-profile event such as the Comair case or the heparin case discussed earlier. The focus is generally on preventing another accident of a similar nature: prevent a wrong runway takeoff or prevent a heparin overdose. Based on the historical data available on the incidents associated with similar causal factors, specific safety performance goals can be established. For example, while it may be desirable to reduce runway incursions to zero, the FAA has established a phased approach. Accordingly, the first performance goal is to reduce runway incursions by 10 per cent by FY2013.[16] This goal has been translated into specific goals for the appropriate business units within the FAA and specific tactics like training, safety equipment installation, and revisions in air traffic procedures or airport markings have been identified.

Similarly, the Joint Commission has established a national goal to 'use medicines safely'. Ascension Health established two clear safety performance goals to reduce adverse drug events across all its hospitals (Butler, Mollo, Gale, and Rapp, 2007):

---

16   FAA 2009-2011 Runway Safety Plan; available at <http://www.faa.gov/airports/runway_safety/>.

- Track hospital-wide improvements as observed by the Institute of Healthcare Improvement's Adverse Drug Event (ADE) Trigger Tool.
- Implement changes to reduce by at least 30 per cent ADEs related to at least one of the four high-risk medication categories: insulin, anticoagulants, narcotics/sedatives, and medication reconciliation. Acknowledging that such a change is incremental, Ascension Health defined initial success as 60 per cent of the acute care hospitals achieving the 30 per cent ADE reduction goal.

Proactive Safety Performance Management Strategy

National-level safety performance goals may be established in both aviation and healthcare and then used to set organizational targets. Further, most of the performance targets could also be expressed in terms of financial impact. In healthcare, e.g. if each VAP infection is estimated to cost the hospital about $15,275, reduction of VAP infections from a national average of 8.2 to 4.1 infections per 1,000 days will amount to a direct cost saving of $62,627 per 1,000 days. At one hospital, Bird, O'Donnell, Silva, Korn, Burke, Burke, and Agarwal (2009) reported a cumulative cost (direct cost plus indirect cost) saving of $1.08 million by reducing the VAP rate from 10.2 to 3.4 per 1,000 days.

In aviation, groups such as the Commercial Aviation Safety Team tend to establish national priorities for safety; however, the financial equivalent of safety performance targets is still under development. In the meantime, the FAA is encouraging airlines, repair stations, airports, and most of its own units to voluntarily implement a Safety Management System (SMS). According to the sample SMS Manual (FAA, 2008), safety risk assessment and financial risk assessment can be combined.

Predictive Safety Performance Management Strategy

Predictive safety performance management involves deep analysis of safety data across multiple safety programmes within one organization and/or across similar safety programmes across multiple organizations. Aviation Safety Information Analysis and Sharing[17] (ASIAS) system is a good example of both types of data analysis. Such sophisticated analysis requires qualitative analysis tools like text analysis integrated with quantitative analysis tools like the FOQA data visualization and analysis software. Latest data fusion and visualization techniques could be used to integrate multiple forms and sources of data to provide a fully integrative illustration of past/current safety gaps as well as predictive simulation of the future.

---

17   Aviation Safety Information Analysis and Sharing, available at <http://www.asias.faa.gov>.

Pardee (2009) demonstrated how multiple, previously isolated databases could be integrated to produce comprehensive and unprecedented results. The key advantage of such analysis is that it enables identification of unique combinations of hazards that could present unprotected risks, accidents waiting to happen. Pardee (2009) illustrates how certain flights were dropping below the minimum vectoring altitude and increasing risk of collision with the terrain. The aircraft traffic tracks illustrate the typical paths that were likely to violate the altitude restrictions. The airport procedures data indicate how the arrival procedures for the particular airport were directing the aircraft along a specific path. The terrain data provides a graphic overlay of topographical conditions under which the flights operate. Finally, the weather data provide additional perspective on how visibility and/or precipitation might impact the minimum altitude violation.

Chapter Summary

In this chapter, we focused on safety performance, the top layer of the Safety Culture Pyramid. Safety performance could be expressed in terms of specific events or error types underlying the events. Further financial impact of the errors could also be expressed in terms of macro-level impact on the organization/industry and micro-level impact on specific types of events. Additionally, impact of errors on the people committing such errors, particularly in the healthcare sector is discussed. Clearly, undesirable events in healthcare and aviation have a profound impact on individuals committing the error, as well as on the organization and the industry. Examples from both aviation and healthcare were discussed.

The accountability scale was used to discuss the impact of the Comair accident on the continued viability of the Aviation Safety Action Program. Comair as well as the aviation industry could have been pushed back from a reporting culture to blame culture. The Comair pilots or most of the aviation industry did not realize that the confidentially submitted pilot reports could be made available to the plaintiffs. From the plaintiff's perspective, it is reasonable to ask for these reports because if the company and/or the FAA knew about the latent systemic failures that contributed to this accident, and they did not address these issues, then their inactions might qualify as negligence. The Patient Safety and Quality Improvement Act of 2005, however, considers patient safety work to be privileged and not subject to Federal, State, or local civil, criminal, or administrative subpoena or order. So, the healthcare industry is not likely to face the Comair-type challenges.

In the heparin case, there appeared to be a blame culture at Cedars-Sinai, but not at the Methodist Hospital. The law suit filed by Dennis Quaid and his wife appeared to be motivated by their desire to pull the high-dosage vials off the shelf and to repackage them to minimize the chance of similar errors in the future. In the United States most hospitals participate in adverse event reporting programmes with their respective Patient Safety Organization, and the Agency for Healthcare Research and Quality is aggregating such reports into a national database; the healthcare industry has moved toward a reporting culture. Many hospitals are also participating in the MEDMARX database to report medication errors. Therefore, non-punitive, voluntary error reporting has become part of the healthcare industry; clearly, the industry has achieved a reporting culture. Further, these databases are likely to have compatible data structures and taxonomies because of the leadership roles played by AHRQ, the Joint Commission, and key hospitals. Soon, the Patient Safety Organizations will be able to release bulletins alerting hospitals and staff of adverse events and errors. Then the industry will have achieved a transformative learning culture.

Finally, integration of qualitative and quantitative data analysis tools could develop unprecedented visualization of error trajectories and identify deeply hidden systemic failures.

References

Aboumatar, H., Blackledge, C., Dickson, J., Heitmiller, E., Freischlag, J. and, Pronovost, P. (2007). A descriptive study of morbidity and mortality conferences and their conformity to medical incident analysis models: results of the morbidity and mortality conference improvement study, phase 1. *American Journal of Medical Quality*, (22), 232–238.

American Medical Association (ed.) (1995). Essentials and information items: accreditation council for graduate medical education. *Graduate Medical Education Directory 1995–1996*.

Bagian, J., Gosbee, J., Lee, C., Williams, L., McKnight, S., and Mannos, D. (2002). The veterans affairs root cause analysis system in action. *Journal on Quality Improvement*, 28(10), 531–545.

Bain, A. (1998). Social defenses against organizational learning. *Human Relations*, 51(3), 413–429.

Battles, J., Dixon, N., Borotkanics, R., Rabin-Fastmen, B., and Kaplan, H. (2006). Sensemaking of patient safety risks and hazards. *Health Services Research*, 41 (4–2), 1555–1575.

Berenholtz, S., Hartsell, T., and Pronovost, P. (2009). Learning from defects to enhance morbidity and mortality conferences. *American Journal of Medical Quality*, 24(3), 192–195.

Berriel-Cass, D., Adkins, F., Jones, P., and Fakih, M. (2006). Eliminating nosocomial infections at ascension health. *Joint Commission Journal on Quality and Patient Safety*, 32(11), 612–620.

Bird, D., O'Donnell, C., Silva, J., Korn, J., Burke, R., Burke, P., and Agarwal, S. (2009). Adherence to VAP bundle decreases incidence of ventilator-associated pneumonia in the surgical intensive care unit. Presented at the 90th Annual Meeting of the New England Surgical Society. Retrieved from <http://www.nesurgical.org/abstracts/2009/ brief2.cgi> on April 24, 2010.

Bosch, J.-C. Eckard, E., and Singal, V. (1998). The competitive impact of air crashes: stock market evidence. *Journal of Law and Economics*, 41(2), 503–519.

Butler, K., Mollo, P., Gale, J., and Rapp, D. (2007). Eliminating adverse drug events at ascension health. *Joint Commission Journal on Quality and Patient Safety*, 33(9), 527–536.

Carroll, J., Rudolph, J., and Hatakenaka, S. (2002). Lessons learned from non-medical industries: root cause analysis as culture change at a chemical plant. *Quality and Safety in Healthcare*, 11, 266–269.

Christensen, J., Levinson, W., and Dunn, P. (1992). The heart of darkness: the impact of perceived mistakes on physicians. *Journal of General Internal Medicine*, 7, 424–431.

Deis, J., Smith, K., Warren, M., Throop, P., Hickson, G., Joers, B., and Deshpande, J. (2008). Transforming the morbidity and mortality conference into an instrument for system-wide improvement. *Advances in Patient Safety: New Directions and Alternative Approaches*. Rockville: Agency for Healthcare Research and Quality.

Dekker, S. (2007). *Just Culture: Balancing Safety and Accountability*. Aldershot: Ashgate.

Delbanco, T. and Bell, S. (2007). Guilty, afraid, and alone: struggling with medical error. *New England Journal of Medicine*, 357(17), 1682–6183.

Engel, K., Rosenthal, M., and Sutcliffe, K. (2006). Residents' responses to medical error: coping, learning, and change. *Academic Medicine*, 81(1), 86–93.

FAA (2008). Sample SMS manual. Retrieved from <http://www.mitrecaasd.org/SMS/ documents.html> on April 24, 2010.

Fryer-Edwards, K., Van Eaton, E., Goldstein, E., Kimball, H., Veith, R., Pellegrini, C., and Ramsey, P. (2007). Overcoming institutional challenges through continuous professionalism improvement: the University of Washington experience. *Academic Medicine*, 82(11), 1073–1078.

Gallagher, T., Waterman, A., Ebers, A., Fraser, V., and Levinson, W. (2003). Patients' and physicians' attitudes regarding the disclosure of medical errors. *Journal of the American Medical Association*, 289(8), 1001–1007.

Haig, K., Sutton, S., and Whittington, J. (2006). SBAR: a shared mental model for improving communication between clinicians. *Joint Commission Journal on Quality and Patient Safety*, 32(3), 167–175.

Heinrich, H. (1931). *Industrial Accident Prevention: A Scientific Approach*. New York: McGraw-Hill.

Helmreich, R., Klinect, J., and Wilhelm, J. (1999). Models of threat, error, and CRM in flight operations. In: *Proceedings of the Tenth International Symposium on Aviation Psychology* (pp. 677–682). Columbus: The Ohio State University.

Houston, D. and Allt, S. (1997). Psychological distress and error making among junior house officers. *British Journal of Health Psychology*, 2, 141–151.

Jaques, E. (1955). Social systems as a defence against persecutory and depressive anxiety. In: M. Klein, P. Heimann and R.E. Money-Kyrle (eds), *New directions in Psychoanalysis*. London: Tavistock Publications.

Joint Commission Resources (2007). Pushing the envelope on root cause analysis: implementing challenging actions for long-term results. *Joint Commission Perspectives on Patient Safety*, 7(3), 3–14.

Joint Commission Resources (2008). Errors associated with new technology. *Joint Commission Perspectives on Patient Safety*, 7(5), 5–6 and 13–14.

Kravet, S., Howell, E., and Wright, S. (2006). Morbidity and mortality conference, grand rounds, and the ACGME's core competencies. *Journal of General Internal Medicine*, 21, 1192–1194.

Latino, R. and Latino, K. (2006). *Root Cause Analysis: Improving Performance for Bottom-line Results*. Boca Raton: CRC Press/Taylor and Francis.

Lazare, A. (2006). Apology in medical practice: an emerging clinical skill. *Journal of the American Medical Association*, 296(11), 1401–1404.

Leonard, M., Graham, S., and Bonacum, D. (2004). The human factor: the critical importance of effective teamwork and communication in providing safe care. *Quality and Safety in Healthcare*, 13, 85–90.

Marx, D. (2009). *Wack-a-mole: The Price we Pay for Expecting Perfection*. Plano: Your Side Studios.

McCaughey, B. (2008). Unnecessary deaths: the human and financial costs of hospital infections (3rd edition). Committee to Reduce Infection Deaths <http://www.hospitalinfection.org>.

Menzies, I. (1960). A case-study in the functioning of social systems as a defence against anxiety: a report on a study of the nursing service of a general hospital. *Human Relations*, 13(2), 95–121.

Mizrahi, T. (1984). Managing medical mistakes: ideology, insularity and accountability among internists-in-training. *Social Science and Medicine*, 19(2), 135–146.

Newman, M. (1996). The emotional impact of mistakes on family physicians. *Archives of Family Medicine*, 5, 71–75.

Orlander, J., Barber, T., and Fincke, B. (2002). The morbidity and mortality conference: the delicate nature of learning from error. *Academic Medicine*, 77(10), 1001–1006.

Ornstein, C. (2007). Actor files suit in drug mishap. *Los Angeles Times*, December 5. Retrieved from <http://articles.latimes.com/2007/dec/05/local/me-quaid5> on April 24, 2010.

Paradies, M. and Unger, L. (2000). *TapRoot: The System for Root Cause Analysis, Problem Investigation, and Proactive Improvement*. Knoxville: System Improvements.

Pardee, J. (2009). Improving safety through collaborative safety initiatives. Presented to the ICAO Air Navigation Commission. Retrieved from <www.icao.int/anb/safetymanagement/../ICAO per cent20CAST per cent206Oct09_R.ppt> on April 24, 2010.

Patankar, M.S. (2002). Root cause analysis of rule violations by aviation maintenance technicians. Grant No. 2001-G-001 Washington, DC.: FAA Office of Aviation Medicine.

Patankar, M.S. and Mondello, S. (2010). SMS safety investment model. Presented to the Federal Aviation Administration, April 7.

Patankar, M.S. and Taylor, J. (2004). *Applied Human Factors in Aviation Maintenance*. Aldershot: Ashgate.

Pierluissi, E., Fischer, M., Campbell, A., and Landefeld, C. (2003). Discussion of medical errors in morbidity and mortality conferences. *Journal of the American Medical Association*, 290(21), 2838–2842.

Reason, J. (1997). *Managing the Risk of Organizational Accidents*. Aldershot: Ashgate.

Rex, J., Turnbull, J., Allen, S., Vande Voorde, K., and Luther, K. (2000). Systematic root cause analysis of adverse drug events in a tertiary referral hospital. *Joint Commission Journal on Quality Improvement*, 26(10), 563–575.

Shappel, S. and Wiegmann, D. (2001). Unraveling the mystery of general aviation controlled flights into terrain accidents using HFACS. Paper presented at the 11th International Symposium on Aviation Psychology. Columbus: The Ohio State University.

Shendell-Falik, N., Feinson, M., and Mohr, B. (2007). Enhancing patient safety: improving the patient handoff process through appreciative inquiry. *Journal of Nursing Administration*, 37(2), 95–104.

Stewart, K. (2007). Two reports from 1990s forwarded to NTSB. Courier-Journal.com, March 16. Retrieved from <http://www.courier-journal.com/article/20070316/NEWS01/101290018/2-reports-from-90s-forwarded-to-NTSB> on April 24, 2010.

Swing, S. (1998). ACGME launches outcomes assessment project. *Journal of the American Medical Association*, 279(18), 1492.

Szekendi, M., Barnard, C., Creamer, J., and Noskin, G. (2010). Using patient safety morbidity and mortality conferences to promote transparency and a culture of safety. *Joint Commission Journal on Quality and Patient Safety*, 36(1), 3–9.

Taxis, K., Gallivan, S., Barber, N., and Franklin, B.D. (2006). Can the Heinrich ratio be used to predict harm from medication errors? Technical Report. University of Birmingham. Retrieved from <http://eprints.pharmacy.ac.uk/764/> on April 24, 2010.

Vincent, C. (2003). Understanding and responding to adverse events. *The New England Journal of Medicine*, 348(11), 1051–1056.

Vincent, C., Taylor-Adams, S., and Stanhope, N. (1998). Framework for analysing risk and safety in clinical medicine. *British Medical Journal*, 316(7138), 1154–1157.

Wald and Shojania, K. (2001). In: Making healthcare safer: a critical analysis of patient safety practices. Chapter 5. Root cause analysis. Evidence report/technology assessment, No. 43. AHRQ Publication No. 01-E058, July 2001. Rockville: Agency for Healthcare Research and Quality. <http://www.ahrq.gov/clinic/ptsafety/>.

Weiss, K., Leveson, N., Lundqvist, K., Farid, N., and Stringfellow, M. An analysis of causation in aerospace accidents. Paper presented at Space 2001. Albuquerque.

West, C., Huschka, M., Novotny, P., Sloan, J., Kolars, J., Habermann, T., and Shanafelt, T. (2006). Association of perceived medical errors with resident distress and empathy: a prospective longitudinal study. *Journal of the American Medical Association*, 296(9), 1071–1078.

Wilf-Miron, R., Lewenhoff, I., Benyamini, Z., and Aviram, A. (2003). From aviation safety to medicine: applying concepts of aviation safety to risk management in ambulatory care. *Quality and Safety in Healthcare*, 12, 35–39.

Wilson, P., Dell, L., and Anderson, G. (1993). *Root Cause Analysis: A Tool for Total Quality Management*. Milwaukee: American Society for Quality Press.

Woolf, S., Kuzel, A., Dovey, S., and Phillips, R. (2004). A string of mistakes: the importance of cascade analysis in describing, counting, and preventing medical errors. *Annals of Family Medicine*, 2(4), 317–326.

Wu, A., Lipshutz, A., and Pronovost, P. (2008). Effectiveness and efficiency of root cause analysis in medicine. *Journal of the American Medical Association*, 299(6), 685–687.

Chapter 4

Safety Climate

Introduction

Safety Climate is the next layer of the Safety Culture Pyramid. Survey questionnaires are commonly used to measure safety climate, which is a snapshot of the attitudes and opinions of the sample population at the time of the survey. There are over 50 safety culture/climate survey instruments! So, in this chapter, we will discuss how safety climate questionnaires are developed, how they are tested for reliability and validity, and how they are used to assess the current state of safety climate as well as effects of specific interventions such as training on safety climate. Next, we will discuss the Patankar and Sabin Safety Climate Questionnaire (PSSCQ), its psychometric properties, and the evolving influence structures of various factors contained in this questionnaire.

Basics of Safety Climate Measurement

Safety climate measurement typically starts with a combination of interviews, focus group discussions, and ethnographic observations. Collectively, these experiences provide a first-level impression about the safety climate at a particular organization at the time of these experiences.

Interviews and Focus Group Discussions in ATO Technical Operations

Our Air Traffic Organization (ATO) Technical Operations (Tech Ops) Safety Culture Research started with a simple question from the vice president: what is the safety culture like in Tech Ops? At that point, the term *safety culture* was not fully understood, but it was gaining popularity across the ATO. So, we developed a set of questions that would facilitate a conversation with individuals and/or small groups. Since Tech Ops is a national organization, which is divided into three major service areas (western, central, and eastern), we sought permission to interview individuals who represented a sufficiently inclusive sample of job titles and technical specialties across each service area.

Based on our review of safety culture literature (Patankar et al. 2005) and one-on-one interviews with key personnel from Tech Ops, we developed the following list of questions/discussion items for our focus groups. However, the most important aspect of this discussion was to start a conversation and to let the people talk about their experience and their opinions about safety culture. In total, over 75 individuals participated in these discussions.

- What's the first word/phrase that comes to mind when you hear 'Safety Culture'?
- How would you describe the general work environment in your unit/department at this time?
- What would you consider a 'success story' that illustrates a strong positive safety culture (this could be in your organization or somewhere else)?
- What are some of the key reasons to improve the safety culture?
- What is your experience with getting some systemic improvements accomplished?
- What are some of the necessary steps in improving the safety culture in your department/location?
- What are some of the challenges in accomplishing the above steps?
- What would you suggest to make the above changes last even when some of the key people have moved on to other positions?
- How would you describe the employee-management relationship with regard to mutual trust, interpersonal communication, professionalism, goal sharing?
- What are the key performance parameters on which you are evaluated and your reward/penalties are decided?
- How do you compare/contrast the safety culture in private industry – whether it is in aviation or outside – with that in your organization? Are there any best practices from these external organizations that could be customized and incorporated in your organization?
- How would the employees characterize their managers in this organization? How would the managers characterize their employees in this organization? How would you rate your organizational unit with your peer units – worse than others, about the same as others, better than others?
- Are there any policy changes that need to be implemented in order to improve the safety culture?
- What type of support do you need from your management/headquarters to improve the safety culture?
- What type of support do you need from your labour union to improve the safety culture?
- How long would it take to change the safety culture in your organization?
- How would you know when you have succeeded in accomplishing the cultural change? How would you prevent a relapse?
- Would you like to add any questions?

All focus-group discussions were audio recorded and transcribed. Content analysis was performed to identify key themes and the reliability of the theme extraction was checked by three independent researchers. The following sections present the results of this analysis.

Perceptions regarding 'safety culture' Overall, the term *Safety Culture* was not familiar to most Tech Ops personnel; some related it to personal injury issues, some to the National Airspace System (NAS) safety, and others simply considered it to be just another buzz word. Some examples of work norms revealed that the field personnel routinely engaged in 'workarounds', which are maintenance actions that are not strictly in accordance with the published procedures. In most cases, the personnel developed these workarounds because there were no published procedures for the particular maintenance tasks. Another issue raised, particularly by experienced personnel, was about perceived overemphasis on minimizing cost. In their opinion, the message of efficiency was so strong that the expectation of safety was getting drowned. They questioned whether the long-held goal of ATO – to provide 'safe, orderly, and expeditious flow of air traffic' – was being impacted by increasing priority being placed on minimizing costs and expediting air traffic. This concern was strong enough that those who were familiar with the *Columbia Accident Investigation Board Report* commented that the FAA's safety culture might be headed toward the NASA Culture of 'faster, better, cheaper' (CAIB, 2003) – implying that a certain degree of deterioration in values might be underway. Nonetheless, it was clear in these discussions that Tech Ops is extremely dedicated to keeping the NAS safe. The people had a 'can do' attitude, but the ability to maintain the safety of the system was expected to reach a critical point in the near future with the increasing number of aging 'unreliable' technical systems that continued to be in service. Coupled with the shrinking workforce, the expectation of high reliability was becoming a 'mission impossible'.

State of the current work environment Tech Ops employees took great pride in their work and exhibited 'ownership' of the equipment for which they were responsible; they strongly believed that they played an important role in ensuring safety of the NAS; they often went beyond their call of duty to minimize errors; and they exhibited ingenuity in developing field-level solutions to a variety of problems. There seemed to be a reasonably positive work environment at the local level; the relationship between the local, regional, and headquarters, however, appeared to be increasingly disjointed. Some participants complained about serious communication disconnects between the local, regional, and headquarters levels, which had resulted in local improvisations to systemic problems. There was a marked discontent regarding the decreasing staffing levels and a concern that such cuts in personnel might affect safety; a few examples were provided to illustrate this point. Cases of training deficiencies were common and were linked with staffing shortages and aggressive equipment deployment schedules. Even in some of the 'better' facilities, the employee-management trust was questionable. One of the participants offered the following comment: 'if errors/problems are reported, the system-wide version that is disseminated will often be sanitized such that critical details are removed for political reasons to protect management'. Managers complained about having limited resources to manage their employees and expressed concern that the message of efficiency was stronger than that of safety.

Success stories In the Tech Ops world, safety of operation is equated with hardware reliability and availability. If the navigation systems, communication systems, radar units, and the environmental systems are operating at or above the target reliability and availability threshold, people regarded the system to be operating at a safe level. Therefore, they regarded each day of safe operation as a 'success story' for the system. When pressed for issues more directly related to safety, the participants described the 'Lessons Learned' database, which is used to document an incident or an accident and the lessons learned from such an event. This database appeared to be a one-way repository of documents that were not used for broader education or development of system-wide solutions. Further, some employees regarded the process of writing a lessons learned report to be punitive in itself. Another database, the Tech Net database, was perceived more positively. It was a database of technical problems and the corresponding solutions. The technical specialists were very comfortable using the Tech Net database to log solutions to problems experienced by others as well as to seek solutions to their own technical challenges.

Key reasons to improve the current safety climate/culture Initially, most participants commented that their current safety culture was very good and needed little improvement. Their reasoning was based on a high equipment availability level: 97–99 per cent. However, as the discussion continued, they acknowledged that the number of highly skilled senior staff is decreasing (due to retirement) while the number of new systems to be maintained and the number of flights are increasing; the system is being stressed. The safety limits of this system are not known, but it seemed to be held together by several individuals who routinely went beyond their call of duty to ensure systemic safety. Clearly, systemic solutions are needed to not only improve the current level of safety, but even to simply maintain the current level of safety.

Experience with getting systemic improvements accomplished There were many examples of local improvements/best practices, but very few that had a national impact. For example, one person rewrote the certification exams for Display System Replacement (DSR) and had them published nationwide – he was very satisfied with the outcome. Another example is when a security planning group worked with First Responders, FBI, and Secret Service to develop a plan to handle terrorism and national emergencies. Overall, many employees had participated in planning for improvements, but their recommendations were not implemented; hence, they are now reluctant to participate in similar efforts. It is generally difficult to make systemic improvements due to organizational bureaucracy; information exchange is typically not effective.

Challenges to accomplishing improvements Since Tech Ops is a national organization, it is often difficult to contact people across time zones. Tension between management and employees seemed to be high in some locations, generally due to mistrust resulting from performance and budget issues. The decline in staffing levels was believed to be reducing redundancies and impacting safety of individuals as well as the NAS. The following examples were shared to illustrate that the risk tolerance is on the rise:

- One person is responsible for maintaining two geographically separated systems. If both these systems fail at the same time, one person will not be able to restore both systems simultaneously; he will have to choose one system over the other and the one that remains out of service will pose a continued risk to the system.
- Instead of two, only one person is sent out on a job at remote sites. If that person gets electrocuted, there is nobody to call for assistance; also, if that person makes a mistake, there is nobody to double-check.
- Many-a-times, there is only one person on the watch for critical systems at night. Sometimes, that person does not have the training or the experience to address the wide range of calls that he/she might receive. There is no back-up.

From a management perspective, tension between production and safety is real and better training is needed to manage this tension. Poor communication between local, regional, and headquarters levels inhibits field problems from reaching appropriate people higher in the management for timely attention. Infrequent personal contact between the three levels breeds misunderstandings and sends mixed messages.

Employee-management relationship Most people were satisfied with the employee-management relationship at the local level. However, some sites appeared to have very low employee-management trust. Across the board, many employees seem to be threatened with the increasing emphasis on job reductions. Some managers seem to be unprepared to handle declining staffing and increasing workloads, particularly with existing communication challenges between local, regional and national levels, as well as the ineffective employee and management evaluation systems. Managers with prior Tech Ops field experience or better communication with the workforce were more respected by their employees. Several people criticized the headquarters due to a lack of personal contact and believed that the problems they faced in fielding new systems could have been averted by headquarters. A common theme expressed was that words speak of safety and actions speak of efficiency: 'HQ does not practice what it preaches'.

Parameters for performance evaluation Both managers and employees were dissatisfied with their respective evaluation systems because they were not always based on merit, there were no clear and efficient means to handle substandard performance, and different managers handled evaluations differently. From a safety perspective, it was interesting to note that nobody was evaluated on safety. Different locations/sites were being held to different performance standards. Many expressed the belief that the annual evaluation was a joke, a perfunctory five-minute interaction. Further, there was no clear career progression ladder in the system or no incentive to move from a less complex facility to a more complex facility or to the headquarters.

Comparison with external organizations or private industry Some locations rated themselves better than private industry; others rated themselves worse than private industry. Many individuals indicated that the US military had better performance standards regarding safety.

Necessary policy-level changes Generally, people believed that policies would not change until an accident occurred. Employees believed that Tech Ops was very reactive: once people are killed, more staff would be hired. According to them, in order to change the safety culture, Tech Ops needs to be more proactive. For example, the new maintenance philosophy, which is a shift toward reliability-centered maintenance, relies on system service data that may not be accurate. In the past, technical specialists often performed maintenance actions even if they were not scheduled to perform the maintenance because they were either in the vicinity of the equipment or noticed some anomaly that required attention. Many technical specialists believed that such spontaneous maintenance actions tend to provide a false sense of reliability in aging equipment. As the organization moves toward condition monitoring and extended maintenance intervals, the safety risk may increase. A real risk-assessment and associated guidance material is needed to assist in performance vs. safety decisions at the operational level.

Need for management support Field personnel felt that the headquarters was getting increasingly out of touch with the local issues. Some managers were accused of changing outage reports to make their facility look better. For example, 'An unscheduled outage may be changed to a scheduled outage to make the facility look better; however, the result is that equipment appears not to need preventative maintenance because it did not fail. If equipment is not prone to periodic failure, then less manpower is needed to maintain it and new hires are not made. This creates a culture based on lies.' For the culture to change, problem managers and supervisors must be changed. Managers need to better understand the technical needs rather than focus on manipulating the maintenance intervals to match the staffing/budget limitations.

Need for labour union support Some managers report that they would like to see the union focus more on organizational (FAA) goals rather than on individual member's needs. Some employees think that the union (PASS – Professional Aviation Safety Specialists) is too weak – not like NATCA (the National Air Traffic Controllers Association). Some employees joined the union to protect their jobs; others were active in influencing local changes and improving work conditions. Some employees were satisfied with the local union representatives, but not with the national representatives. PASS was believed to be frequently reactive to management instead of being proactive and a strong, equal partner in key decisions.

Comments from open discussion Aggressive technology/system implementation schedule undermines quality/safety. FAA does not want to hear grass roots problems. People who maintain the systems have little opportunity to provide input to the design and acquisition of such systems. Great performers are not recognized; poor performers are not penalized. Concern regarding the overall shift in who the customer is: from the flying public to the airlines. While senior staff loved their job in the past, they are now looking forward to retirement – it's no longer worth the effort.

Important issues across the system The focus group discussions can be further distilled into seven key themes: (a) identity crisis, (b) interpersonal communication, (c) labour-management relationships, (d) interpersonal trust, (e) goal sharing/ attainment, (f) performance standards and evaluation, and (g) employee involvement in shaping the collective future. Essentially, the Tech Ops employees are talented professionals committed to their mission of ensuring NAS safety. At the time of the focus group discussions, they indicated some confusion about their organizational priorities and hence there is some question regarding their identity – is safety an organizational priority over efficiency? Are they working toward a safer system for the flying public or are they working toward a more efficient system for the airlines? The issues with interpersonal communication, labour-management relationships, and interpersonal trust are rather typical of a large and distributed organization like Tech Ops; however, a unifying safety philosophy coupled with appropriate training materials would go a long way in improving communication and interpersonal relationships. Changes to performance evaluation systems and opportunities for employee involvement are critical in changing the organizational culture in the future. These two items are the classic artefacts of the old culture that will keep resisting adoption of the new culture.

Development of the Patankar and Sabin Safety Climate Questionnaire

The focus group discussions presented earlier and the literature review discussed in this section led to the development of the Patankar and Sabin Safety Climate

Questionnaire (Patankar and Sabin, 2008). This questionnaire has three sets of factors: organizational factors, team factors, and outcome factors. The specific items that relate to each of the factors are presented.

Organizational Factors

Knowles (2002) claims institutional identity, information flow, and relationships as the three key aspects for managing organizational success. He demonstrated through several case examples how he was able to improve both safety as well as business performance at DuPont chemical plants using his Process Enneagram. Based on additional literature in safety (cf. Westrum, 1995; Reason, 1997; Patankar and Taylor, 2004) and organizational change management (cf. Collins and Porras, 1997; Wheatley, 1999; Collins, 2001), Patankar et al. (2005) added leadership and evaluation (cf. Roughton and Mercurio, 2002) as two other core elements necessary for a successful safety management system.

Institutional identity   Understanding institutional identity is critical to sustaining new programmes – whether they are safety programmes or business initiatives. Collins and Porras (1997) studied several companies and deduced that successful companies were better able to 'preserve the core and stimulate progress'. In reviewing the progress of Maintenance Resource Management Programs, Patankar and Taylor (1999) note that the absence of an umbrella document that connected these programmes with the institutional goals may have at least contributed to the deterioration of such programmes. When a programme is tied with the core corporate identity and goals, it receives attention from the highest level of the corporate management (Knowles, 2002). Consequently, everyone down the chain of command is more likely to be held accountable for the success of such programmes. For example, when DuPont Chemicals realized that their business success was fundamentally dependent upon their safety programmes, a culture of high reliability (and high safety) emerged. Similarly, when the DuPont management recognized that it was essential to collaborate with their stakeholders, the community in which DuPont's plants were located and their regulators, they acknowledged that everyone's survival was interdependent, and they developed comprehensive safety programmes (Knowles, 2002). There are many such similarities in the nuclear power sector (cf. Schulman, 1993) as well as in other high-reliability sectors (cf. Roberts, 1993). The basic fact is that in organizations with a strong safety culture, safety is integral to the organization's survival/success.

Sloat (1996) claims, 'an organization is most effective when artefacts, values, and assumptions are lined up and mutually supportive' (p. 66). If such alignment is not orchestrated, the safety programme (or change programme in general) tends to be launched at the artifact level alone – new stickers, logos, uniforms, etc. Eventually, the new initiative 'will be overwhelmed by the practices that are supported by the values and assumptions of the organization' (p. 68). and it will die. Therefore, in order to have a sustainable safety programme, it has to be aligned with the fundamental identity, values, and assumptions of the organization. The espoused safety values may be accessible through corporate promotional materials, but the enacted safety values are usually discoverable through periodic safety climate surveys.

The following questionnaire items are aimed at institutional identity:

- NAS [National Airspace System] safety is consistently communicated as the core value/goal of the ATO-Tech Ops.
- There is high consistency between words and actions throughout ATO-Tech Ops.

Information flow Westrum (1993) views information flow as such a critical element of an organization that he classifies organizations based on this key parameter: pathological, bureaucratic, or generative. Clearly, the ability of an organization to receive good information and act on it in appropriate manner is dependent on the mechanisms that are set up for such information flow. There are many similarities between Westrum's classification and the one by Senge (1990) on learning organizations. Senge views information as key for organizational learning. Based on the growing literature on reporting cultures (Marx, 1997; Reason, 1997; Taylor, 2004; Patankar and Gomez, 2005), it is clear that an organization could transition from a blame-ridden culture to the one that has a 'Just Culture' (Marx, 1997; 2001), if it is successful in implementing an effective reporting culture. The effectiveness of a reporting culture depends on the employees' confidence in the system or the leadership – they need to feel that their feedback/input will be taken seriously (Harper and Helmreich, 2003). The quality of information flow vertically across the organizational ranks, and more importantly the effects of such information (e.g. does anyone really act on this information?) impacts interpersonal trust. Patankar, Taylor, and Goglia (2002) studied interpersonal trust among aircraft mechanics and their supervisors; they report that up to 30 per cent of the mechanics do not trust their supervisors to act in the interest of safety. Later, Patankar, and Driscoll (2004) noted that organizations with error reporting programmes such as the Aviation Safety Action Program had a higher level of interpersonal trust than those without. Therefore, it is not only important to open information flow vertically as well as laterally throughout the organization, but it is also important to act on the information that is received as a result of such open communication.

In an example from the Technical Operations domain of the FAA's Air Traffic Organization, Ahlstrom and Hartman (2001) discovered that according to some specialists (technicians), the equipment status information and information in other databases are not always maintained and up-to-date. This discrepancy can cause errors such as calling a field technician who is unavailable to fix a problem and thus increasing outage durations. The specialists also indicated that weather played a critical factor in Airway Facilities (now Technical Operations) decision-making. However, Ahlstrom and Hartman's observations and structured interviews revealed that specialists often do not have current weather information for their area.

Another source of errors among the specialists was procedural ambiguity or noncompliance. This may be due to lack of training on the part of the specialist or memory overload. These errors occur in the present Maintenance Control Centers (about 40 are strategically placed throughout the United States), but there is also the potential for increased human error of this type with the introduction of new procedures and business practices associated with the Operations Control Centers (OCC) – three centrally located regional centres responsible for monitoring and controlling the facilities in their region, assigning personnel and resources, and coordinating Airway Facilities (now Technical Operations) and Air Traffic Information (Ahlstrom and Hartman, 2001).

The following questionnaire items are aimed at information flow:

- Tech Ops managers will act on NAS safety concerns voiced by their employees.
- Very good communication about NAS safety exists up and down the Tech Ops chain of command.
- Very good dialog and communication exists between Tech Ops and their Enroute and Terminal Air Traffic partners to ensure NAS safety.

Relationships Fundamentally, people make the organization. Hence, it is important to build strong, positive, professional relationships between the people who work together. Knowles (2002) has illustrated in several examples how people come together in crises, forget their individual differences, and work toward the larger problem at hand. If such collaboration and cooperation could be retained after the crisis is over, the organization would function under the 'self-organizing' or 'living system' paradigm. Relationships within the organization are shaped by the level of shared purpose, open communication between the people, and their interdependence. As Knowles experienced, once the information was openly shared and the goals were clearly communicated, the health of the professional relationships among the people within his organization, as well as throughout the larger community in which the organization was located, improved dramatically. From a living systems perspective, positive information flow is bound to result in healthier professional relationships because organizations are living systems and information is the lifeblood of such systems (Wheatley, 1999). It is up to the leadership to take advantage of such natural forces and align them in a positive direction. There is additional literature on the quality of professional relationships, their effect on job satisfaction, and the overall organizational success (cf. Freiberg and Freiberg, 1996; Herzberg, 1966). Therefore, it is important to invest in building strong, professional relationships in an organization.

The following questionnaire items are aimed at relationships:

- There is excellent cooperation between different Tech Ops specialties/disciplines.
- Working in Tech Ops is like being part of a large family.

Leadership Effective leadership appears as a key contributor to successful safety programmes in a variety of domains. Fleming and Lardner (1999) studied the impact of supervisors' attitudes, management style, and behaviour on their subordinates' safety behaviour in the offshore oil industry. This study identified a number of supervisor attributes that were associated with positive subordinate safety behaviour, and less risk-taking behaviour indicated that their supervisor possessed attitudes, skills, and behaviours that can be summarized as follows:

- Valuing their subordinates.
- Visiting the work-site frequently.
- Facilitating work group participation in decision-making.
- Effective safety communication.

Fleming and Lardner's research suggests that a supervisor safety management development programme could be an effective mechanism for safety culture improvement. The factors to be considered when developing a supervisor safety development programme are as follows:

- Supervisor training should include a focus on the interpersonal aspects of safety management.
- Training should be skill-based (the how) as opposed to purely knowledge-based (the what).
- Subordinates should be involved in decision-making.
- A role model should be provided to motivate supervisors and keep the process moving.
- Support should be given from senior and middle management.

In a survey of train drivers in the United Kingdom, Clarke (1998) found that very few drivers (3 per cent) reported other drivers' rule breaking behaviours where a third of the drivers felt that rule breaking by another driver was not worth reporting. Clarke also found that train drivers in the United Kingdom were less likely to report incidents if they considered managers would not be concerned with such reports. High levels of non-reporting were most evident when workers felt that incidents were just 'part of the day's work' and that 'nothing would get done'. These findings indicate that incidents are not reported because they are accepted as the norm; this was further reinforced when drivers perceived that reporting an incident would not result in any action being taken, indicating a lack of commitment by management. However, the results also indicate that drivers would be more likely to report an incident if they thought something would be done to remedy the situation.

The influence of leadership on the success or longevity of a safety programme was also reported by Taylor and Christensen (1998) and Patankar and Taylor (2004). In both of these studies, it was reported that: (a) most maintenance resource management programmes stalled due to a 'lack of management follow-up', and (b) the awareness of safety issues raised by training programmes tends to erode or even turn into a negative attitude, if such awareness is not reinforced by supporting structural or procedural changes. Management must clearly demonstrate that it is willing to make the changes necessary to sustain the attitudes and behaviours espoused by the training programmes and bridge the gap between 'espoused theory' and 'theory in use' (Argyris and Schön, 1974). The Australian Occupational Health and Safety Commission cites, 'Recurring findings across the studies were the critical role played by senior managers in successful health and safety management systems, and the importance of effective communication, employee involvement and consultation' (Gallagher, Underhill, and Rimmer, 2001, p. 12).

In studies of aircraft mechanics and health care professionals, Patankar, Brown and Treadwell (2005) note that the priorities of frontline workers and supervisors or management personnel are different: frontline personnel tend to be focused on technical aspects of their jobs and managers tend to focus on the fiscal and operational aspects of their jobs. If we apply Wheatley's (1999) 'living systems' model, each person in an organization can be motivated to change by linking his/her individual survival (in the job) to that of his/her organizational unit. Patankar, Brown, and Treadwell's studies suggest that frontline personnel are motivated by their respective duty ethics; therefore, if they are mature enough in their ethical decision-making, they will be more concerned with fulfilling their professional duty (their professional survival is tied to the validity of their professional licensure, which, in turn, is tied to adherence to standard professional practices) than in saving money for their company. Since these studies note that managers tend to be more focused on fiscal or operational aspects (their survival as managers is tied to the fiscal performance/efficacy of their organizational unit), one could deduce that the managers could be more motivated to support the safety programme when such a programme demonstrates a specific fiscal impact.

The following questionnaire items are aimed at leadership:

- Tech Ops managers encourage their employees to report safety discrepancies and human errors without fear of negative repercussions.
- I trust the risk assessments made by Tech Ops leaders regarding NAS safety.

Evaluation/accountability Strebel (1998) noted that employees make three levels of compacts in the workplace: first, the formal compact based on their appointment/employment contract – what do they need to do to retain their job; second, based on psychological aspects – how hard do they really have to work, what reward/recognition will they receive, and will it be worth the effort; and third, based on the social aspects – what is the level of consistency between the company's mission statement and experienced practices or what are the real rules that determine who gets what in this company. The tighter the alignment of the change programme with these three compacts, the better the likelihood of its success. Therefore, if safety performance is important to the organization, it must be measured, all employees and their managers must be held accountable for their actions or inactions, they must receive equitable recognition for their efforts, and the policies and practices must be consistent (cf. Roughton and Mercurio, 2002). In organizations with strong safety cultures, the safety goals of the organization are fully understood by each employee (including the CEO) and each person is held accountable for doing their part in achieving the organizational goal (Grubbe, 2003). Therefore, employee evaluation is a critical part of organizational performance: it is the glue that makes change initiatives 'stick'.

The following questionnaire items are aimed at evaluation or accountability:

- For specialists [technicians], employee performance appraisals are closely linked to the Tech Ops goals of maintaining NAS safety.
- For specialists, good work performance related to NAS safety is rewarded.
- For specialists, poor work performance related to NAS safety is promptly addressed and corrected.

Team Factors

Patankar, Bigda-Peyton, Sabin, Brown, and Kelly (2005) placed Professionalism, Interpersonal Trust, Goal Sharing, Adaptability/Resilience, and Institutional Support in the Team Factors category because these scales are dependent on collaboration among coworkers, employee-management relationship, the level of connection an individual feels with the general direction of the organization, etc. Ultimately, these factors seem to cluster around the notion of teamwork.

Professionalism In multiple longitudinal survey research projects, involving several thousand aviation maintenance personnel, researchers have reported that individual professionalism, which is composed of professional competence as well as self-awareness of vulnerability to human performance limitations, is critical to the development of a safer maintenance environment (or culture) (Taylor, 1995; Taylor and Christensen, 1998; Taylor and Patankar, 2001). Later, Patankar, Brown, and Treadwell (2005) added 'the ability to make sound ethical decisions' to the definition of individual professionalism.

In this chapter we present professionalism as a team factor because the notion of individual technical competence seems to be dependent on the perceived identity of the organization (do we hire only the best technicians? Are we renowned for the best in-house training programmes?), as well as the evaluation systems in place (how do we handle lapses in technical performance/skill?). While one's ability to make ethical decisions may be independent of the organization in which they are employed, the gap between individual ethical standards and those of the organization is likely to be influenced by the quality of information flow, leadership, and the evaluation system. Thus, we believe that the organizational factors that were presented earlier have some influence on individual professionalism.

The following questionnaire items are aimed at professionalism:

- Tech Ops employees use adequate redundancy and backup systems to minimize the probability of systemic failure.
- My work group consistently learns from its successes and failures.
- I am proud to work for Tech Ops.
- My work group is the 'best in business' when it comes to supporting NAS safety.

Interpersonal trust Organizations with a positive safety culture are characterized by 'communications founded on mutual trust, by shared perceptions of the importance of safety, and by the efficacy of preventive measures' (ACSNI, 1993). Further, the importance of interpersonal trust in building a strong safety culture has been noted in several studies (cf. Helmreich, 1999; Patankar, Taylor, and Goglia, 2002; Roughton and Mercurio, 2002; GAIN, 2004; Patankar and Taylor, 2004).

The process of clearly establishing acceptable versus unacceptable behaviour, if done properly in a collaborative environment, brings together different members of an organization that might often have infrequent contact in policy decision-making. This contact, as well as the resulting common understanding of where the lines are drawn for punitive actions, enhances the trust that is at the core of developing Just Culture. In order to combat human errors, we need to change the conditions under which humans work. The effectiveness of countermeasures depends on the willingness of individuals to report their errors, which requires an atmosphere of trust in which people are encouraged to provide essential safety-related information. (GAIN, 2004).

The following questionnaire items are aimed at interpersonal trust:

- My supervisor can be trusted to act in the interest of safety.
- My suggestions about safety would be acted on if I expressed them to my supervisor.
- I feel comfortable going to my supervisor's office to discuss safety problems.
- My supervisor listens to me and cares about my concerns.
- My supervisor trusts me.

Goal sharing La Porte and Consolini (1991) studied High Reliability Organizations (HROs) and reported that such organizations have very clearly identified operational goals and the consensus among the employee groups is unequivocal. Considering that much of the HRO research was concentrated on Naval aircraft carrier operations and nuclear power plants, one could visualize a battery of personnel trained to perform specific tasks to perfection on a routine basis. While one might argue that many other systems depend on goal sharing, three other features distinguish the HROs from the rest: (a) extensive system of internal crosschecks to ensure fail-safe performance, (b) constant training and monitoring to encourage a culture of responsibility and accountability, and (c) a high level of social control by limiting influences from environments external to the organization (Clarke and Short, 1993).

In organizations that are less isolated from their environmental influences or the ones that are more open than HROs, the value of goal sharing becomes even more pronounced. Pierce (2005), himself a manager at Massachusetts General Hospital, acknowledges that, 'culture drives quality and safety', and claims that their articulation of the hospital's vision into a simple, straightforward language has provided them with a 'clear measuring stick for evaluating and directing all that goes on'.

There are many ways to measure corporate performance, such as the Balanced Scorecard (Kaplan and Norton, 1998), but the key is to ensure that what's measured is consistent with the desired goals (Zahlis and Hansen, 2005). It is particularly important to address this issue when safety and performance goals may compete against each other. Drury and Gramopadhye (1991) have reported 'Speed-or-Accuracy Trade-off – SATO' as a key challenge for managers in aviation and other high-consequence industries; typically, safety is encouraged as long as it does not interfere with performance targets; when performance targets are endangered, workarounds and violations of safety practices tend to emerge and they tend to be overlooked by management. Patankar and Taylor (1999) observed a decision-making protocol called Concept Alignment Process (CAP) that was used by a corporate flight department to make decisions related to operations, flight safety, and maintenance safety. This process was effective in aligning the daily tasks with the overall goals of the organization as well as with the safety standards. In order for this process to succeed, managers had to commit to backing off on the performance targets if the risk (safety) was elevated beyond the preset acceptable level. Such active risk mitigation is not likely when clear goals are not established and communicated, processes to enforce adherence to preset performance parameters are not clear, or managers are not empowered to scale back on their performance targets when risk is beyond the acceptable level.

The following questionnaire items are aimed at goal sharing:

- The primary objectives of my work group are highly aligned with key Tech Ops goals.
- I strongly agree with ATO's mission and goals.
- I understand my specific responsibilities to support my work group's goals.

Adaptability/resilience Functional and component redundancies are key aspects of reliability in HROs. It is this level of redundancy that allows the system to collapse its hierarchy under crisis and still maintain the performance reliability (Clarke and Short, 1993). Considering that HROs are typically more protected from external environmental influences, they tend to have financial and human resources to maximize their reliability (safety). Since the survival of the organization (nuclear power plant, chemical plant, aircraft carrier, offshore oil platform, etc.) as a whole depends on safe execution of every mission and timely and safe recovery from minor lapses, tremendous emphasis is placed on systemic as well as task-level safety. When such organizations are plagued with economic pressures and redundancy is compromised, their vulnerability tends to increase; fewer people are loaded with more tasks; the 'coupling' of already complex systems increases; and the 'task loading' tends to render individual personnel incapable of responding to the additional workload imposed by a crisis (VanDrie, 2005).

Declining human resources may be compensated by increased automation, but such technology must be matched with the human operators and maintainers – sociotechnical systems must take into account human and machine reliability issues and strive to develop a joint optimization of their individual strengths (Taylor and Fenton, 1993).

The following questionnaire items are aimed at adaptability or resilience:

- I frequently do more than is expected to protect NAS safety.
- For Tech Ops, NAS safety is heavily dependent on the specialist's work ethic.
- I readily adapt to the challenges of getting my job done.

Institutional support systems Regardless of whether cultural design or change efforts start from the top leadership and permeate down the organizational hierarchy or grow from grass-roots efforts and bubble to the top, institutional support systems are vital for such efforts to flourish. For top-down efforts, the typical challenge is in converting the corporate 'propaganda' into reality – bridging the gap between the 'espoused theory' and the 'theory in use' (Argyris and Schön, 1974). Some organizations have used a participative approach that involves key user groups to influence the espoused culture and develop structures and performance parameters to measure goal attainment. Again, while the traditional HROs may have the luxury of limiting the influences of external environments, most other organizations do not; yet, they must raise their safety performance to the level of HROs and beyond. In order to achieve this goal, purposeful structures and processes developed by dedicated leaders throughout the organization are essential.

Depending on how distant a particular organization is from achieving safety at the HRO level or higher, more or less dramatic changes in the norms, policies, procedures, and practices may be necessary. In order to orchestrate such changes, it may be necessary to provide a variety of new services like employee counselling, safety awareness training, on-the-job training, and management/leadership training; new programmes like an error/hazard reporting programme, process improvement programme, event investigation programme, and risk management; and new ways of recognizing employee contributions. Overall, the people must feel that their organization is willing and able to change in order to meet the established safety goals and that everyone, regardless of position or seniority, is held accountable for their actions (or inactions).

The following questionnaire items are aimed at institutional support systems:

- I have resources to do my job the way it should be done.
- My work group has the necessary number of staff to do the job right.
- Specialists receive adequate technical training to perform their work effectively and maintain NAS safety.

The title of this scale was changed to 'Adequate Resources' to better match the Tech Ops domain.

Outcome Factors

Outcome factors are hypothesized to be the effects of organizational factors and team factors. Also, outcome factors are the most visible or most frequently measured. The specific areas covered under these factors are described below.

Employee satisfaction Typically, employee satisfaction surveys give valuable information about the morale and motivation among the employees, their perceptions of their workplace, their self-worth in their jobs, and the employee-management relationship. These surveys may also indicate specific problem areas such as communication effectiveness, response to specific policies that were implemented or are planned, or more general change efforts underway (cf. Hackworth, et al., 2004).

Herzberg (1966) has identified several workplace satisfiers and dis-satisfiers. Most employee satisfaction surveys tend to build upon Herzberg's early work. One key finding from Herzberg's research is that it may not be possible to increase satisfaction by simply reducing the dissatisfaction; moreover, people will find means to overcome their dissatisfaction if they are particularly satisfied about a particular aspect of their job. For Herzberg, reducing dissatisfaction would not create genuine satisfaction, but growth and engagement in the nature of work itself could potentially lead to increased satisfaction.

Lately, employee satisfaction surveys are being used to rank companies as 'best companies to work for'. Such ranking draws the attention of top management and could be used effectively to institute deep, meaningful changes in the organization. One danger in using employee surveys, like any other survey, is that employees expect that the management will act on their findings; they expect that something will change as a result of the survey. If nothing changes, the employee-management relationship tends to suffer (cf. Patankar and Taylor, 2004). Also, while such surveys may be helpful in knowing the status of employee morale, they can also be very useful in diagnosing specific problems.

The following questionnaire items are aimed at employee satisfaction:

- Tech Ops employees have a high level of job satisfaction.
- I am proud to work for Tech Ops.

Customer satisfaction Customer satisfaction is more commonly used as a diagnostic tool as well as a benchmarking tool. Organizations that excel at customer service/satisfaction are often quoted as the ones that are fanatic about customer satisfaction and frontline employees are rewarded/corrected based on the quality of their interaction with their customers (Collins and Porras, 1997, Ch.6). Standardized customer satisfaction measurement tools can be used to identify customer-service problems throughout the organization. Such standardized measurement tools are also effective in communicating the customer satisfaction goals in a clear and consistent manner throughout the organization, especially when the organization is dispersed across the country and has many different technical/functional units.

The following questionnaire item is aimed at institutional identity customer satisfaction:

- En Route and Terminal Air Traffic partners are highly satisfied with Tech Ops work to maintain the reliability and technical safety of the NAS.

Public image/perception Airlines are typically ranked based on criteria such as customer satisfaction, lost/damaged baggage statistics, on-time arrivals and departures, and in-flight service quality (Bowen and Headley, 2002). Similarly, other companies may be ranked or recognized based on certain industry-accepted performance criteria: ISO certification, Malcolm Baldrige National Quality Award, J.D. Power, and Associates Award, etc. While these awards are symbols of recognition by peer organizations, accident/incident reports or large-scale environmental damage reports can severely harm the organization's reputation in the business as well as in the social community in which it operates. It would be valuable to scientifically determine whether or not any of the organizational factors or team factors presented in this report influence any of the public image criteria.

The following questionnaire items are aimed at public image/perception:

- My work group is the 'best in the business' when it comes to supporting NAS safety.
- Safety of NAS has never been better.

The title of this scale was changed to 'Safety Performance' to better match the Tech Ops domain.

Regulatory compliance Aviation, being one of the most highly regulated industries, is most vulnerable to regulatory violations. Federally certificated job functions such as pilot, mechanic, and dispatcher are highly procedural and these procedures are incorporated into specific regulatory requirements. Therefore, violation of any of the prescribed procedures tends to translate into regulatory violation. Since the procedures are intended to provide for a safe execution of the corresponding task, violation of the associated procedure might be an indication of increased risk or reduced safety – most maintenance errors are attributed to procedural violations (Patankar, 2002). Therefore, the number of regulatory violations assessed against a particular aviation company could be regarded as an inverse measure of its safety record.

Errors/violations, whether regulatory or not, tend to be reportable events in most safety-conscious industries. Therefore, these events get investigated; the focus of such investigations is now shifting from the traditional 'blame game' to a more systemic solution. The frequency of errors and more importantly the comprehensive solutions resulting from the investigations could serve as valuable measures of safety culture.

The following questionnaire items are aimed at regulatory compliance:

- Specialists often need to do 'workarounds' to compensate for lack of component compatibility.
- Specialists often receive equipment to be installed and maintained before it is adequately tested.
- Often there are inadequate specifications for compatibility standards between various types of equipment.
- Specialists often need to do 'workarounds' to compensate for inadequate or out of date procedures.

The title of this scale was changed to 'Technical Standards' to better match the Tech Ops domain.

Stakeholder value From a business perspective, many progressive organizations now believe that employees are the primary stakeholders and customers are secondary; take care of the employees and the employees will take care of the customers. Employee satisfaction also impacts staff turnover and the expenses related to re-training. It is important to balance the needs of the internal employees with those of the external business partners and customers. Chemical companies like DuPont have demonstrated outstanding community partnerships that strengthen not only the company's image as a safety-conscious organization, but also the community's trust in the organization as a positive work environment and as a positive influence on the local economy. Similarly, the airline industry has formed a Commercial Aviation Safety Team (CAST) to evaluate and recommend specific safety enhancements. These recommendations are better received by the regulators and the flying public because they are industry-driven and developed in a spirit of partnership. Ultimately, the success of the organization is reflected in the success of its stakeholders: if the airline succeeds/survives, all the associated support businesses thrive/survive; the same is true for other large companies.

The following questionnaire items are aimed at stakeholder value:

- Tech Ops employees have a high level of job satisfaction.
- I am proud to work for Tech Ops.
- En Route and Terminal Air Traffic partners are highly satisfied with Tech Ops work to maintain the reliability and technical safety of the NAS.
- Safety of the NAS has never been better.

The Patankar and Sabin Safety Climate Questionnaire

In the preceding section, we showed you how items for a survey questionnaire could be developed. This approach was used to develop the Patankar and Sabin Safety Climate Questionnaire, which now consists of 58 items. All items are to be scored on a five-point Likert-type scale where one equals strongly disagree, two equals disagree, three equals neutral, four equals agree, and five equals strongly agree. This questionnaire was prepared for the Technical Operations group of the Air Traffic Organization (ATO), FAA.

As you will note, many of the questionnaire items are customized to the Tech Ops domain. However, most of them can be adapted to the language of other industries. For example, the following item could be restated to match the respective industry:
Original:

- NAS safety is consistently communicated as a core value/goal of the ATO-Tech Ops.

Restated for patient safety:

- Patient safety is consistently communicated as a core value of our hospital.

Restated for employee safety:

- Employee safety is consistently communicated as a core value of our company.

Survey Items and Scales

A confirmatory factory analysis was conducted using conceptual assignment of the survey items to the corresponding scale. Based on the reliability analysis of each scale, the item composition was modified. Table 4.1 lists the scales, their corresponding Cronbach's Alpha, a measure of reliability of the scale, account of variance percentage, and Kaiser-Meyer-Olkin (KMO) scores. A minimum Alpha of .500, a minimum per cent variance of 50, and a minimum KMO of .500 were used to accept the scale as reliable and the factor as appropriately loaded.

Each item within each scale had a factor loading of at least 0.500. Following the confirmatory factor analysis, an exploratory factor analysis was conducted. Two new scales emerged: (a) Learning from Errors, and (b) Error Reporting. The original Support Systems scale was renamed Adequate Resources, the Public image/Perception scale was renamed Safety Performance, and the Regulatory Compliance scale was renamed to Technical Standards to better match the domain of this research. All the 'R' items in Table 4.1 are reverse-scored items.

In order to further test, validate, and revise the scales, the questionnaire needs to be administered multiple times – to additional populations (even across multiple professions in the aviation industry as well as across other industries) and to the same population (repeated studies over a period of time to see how the scale holds its reliability).

Table 4.1   Scale composition and reliability

Scale	Items	Cronbach's Alpha	Accounts for Variance (per cent)	KMO
1. Organizational Factors				
1.1 Institutional Identity	1,2	0.775	82	.642
1.2 Information Flow	5,7,8	0.710	64	.638
1.3 Relationships	9,10	0.699	77	.500
1.4 Leadership	11,12	0.777	82	.500
1.5 Evaluation	13,14,15	0.723	65	.668
1.6 Learning from Errors	19–21,24	0.817	65	.801
2. Team Factors				
2.1 Professionalism	33,36,53,56	0.683	51	.687
2.2 Supervisor Trust	25–29	0.948	83	.902
2.3 Goal Sharing	30–32	0.702	63	.658
2.4 Adaptability/Resilience	34,35,37	0.671	61	.664
2.5 Adequate Resources	41–43	0.675	61	.651
2.6 Error Reporting	16R,22,23	0.545	55	.629
3. Outcomes Factors				
3.1 Employee Satisfaction	52,53	0.675	76	.500
3.2 Customer Satisfaction	54	NA	NA	NA
3.3 Safety Performance	56,58	0.661	75	.500
3.4 Technical Standards	48–51R	0.848	70	.755
3.5 Stakeholder Value	52–54,58	0.787	62	.775

Structural Tests

The three-factor structure (organizational factors, team factors, and outcome factors) outlined in the Patankar and Sabin Safety Climate Questionnaire can be tested to determine the influence of one factor over the other(s). Patankar, Bigda-Peyton, Sabin, Brown, and Kelly (2005) presented a conceptual model called the Purpose-Alignment-Control (PAC) Model, which used the above mentioned three-factor structure to illustrate the potential influence of organizational factors on team factors and that of organizational and team factors on the outcome factors. Subsequently, Block (2008) used Structural Equation Modelling (SEM) to test the PAC Model. Block used three fit indices: goodness of fit (GFI), root mean square error of approximation (RMSEA), and comparative fit index (CFI). Using these three fit indices, researchers test a variety of combinations of relationships among factors to ultimately determine the best model fit. Typically, GFI and CFI values above 0.90 and RMSEA value below 0.08 are considered good model fit. Additionally, path coefficients are used to indicate the strength of the relationship between two constructs; the closer the value to 1.0, the stronger the relationship. Overall, the CFI, GFI, and the RMSEA indicate whether the relative relationship of the various factors can be validated with empirical data. The path coefficients indicate the strength of the relationship, indicating how likely it is that one factor might influence the other.

Figure 4.1 illustrates the conceptual PAC Model. Block (2008) supported the general structure of the PAC model, but modified the factor structure to be more consistent with the MRM/TOQ attitude areas (c.f. Taylor, 2000). The GFI for this modified structure was .92, the RMSEA was 0.047 and the CFI was 0.99.

These results suggest that the PAC model could be used to continue to test the relationship between organizational factors, team factors and outcome factors. Also, influence of specific factors on other factors could be studied and appropriate interventions could be developed.

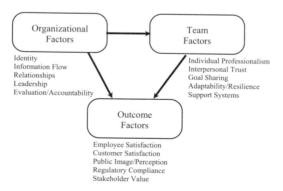

Figure 4.1   Conceptual illustration of factor influences in the PAC model

Chapter Summary

Since there are over 50 different safety climate survey questionnaires, it is not practical to present a comparative review. Instead, this chapter presented a detailed look at development and testing of one safety climate survey questionnaire. The process and the outcomes originated in a study conducted for the Technical Operations domain of the Air Traffic Organization.

Generally, a safety climate survey design starts with a series of focus-group discussions or interviews with key stakeholder groups. Based on the outcome of these discussions as well as the relevant literature, a survey questionnaire can be developed. If the items used in the questionnaire have been validated by other studies, they have a better chance of demonstrating validity and reliability. New and/or unique items could be tested for validity and reliability using accepted statistical methods. Ultimately, a structural equation modeling approach may be used to describe the influence of one group of factors over others.

In the example used in this chapter, a three-factor structural model emerged. Organizational factors are believed to influence team factors, and organizational factors and team factors are believed to influence outcome factors.

References

ACSNI (1993). *ACSNI human factors study group. Third report: organising for safety, advisory committee on the safety of nuclear installations, health and safety commission.*

Ahlstrom, V. and Hartman, D. (2001). Human error in airway facilities. Technical report DOT/FAA/CT-TN01/02. Retrieved from <http://www.hf.faa.gov/> on November 14, 2005.

Argyris, C. and Schön, D. (1974). *Theory in Practice: Increasing Professional Effectiveness.* San Francisco: Jossey-Bass.

Block, E. (2008). Maintenance resource management in aviation: a systems approach to measuring training impact. Doctoral dissertation. St. Louis: Department of Psychology, Saint Louis University.

Bowen, B. and Headley, D. (2002). The airline quality rating 2002. Report by the University of Nebraska at Omaha and Wichita State University. Retrieved from <http://webs.wichita.edu/depttools/DeptToolsMemberFiles/aqr/aqr.pdf> on November 14, 2005.

CAIB (2003). Report of the Columbia accident investigation board, volume 1. Retrieved from <http://caib.nasa.gov/news/report/volume1/ default.html> on August 30, 2005.

Clarke, L. and Short, J. (1993). Social organization and risk: some current controversies. *Annual Review of Sociology*, 19, 375–399.

Clarke, S. (1998). Organizational factors affecting the incident reporting of train drivers. *Work and Stress*, 12, 6–16.

Collins, J.C. (2001). *Good to Great: Why Some Companies Make the Leap ... and Others Don't*. New York: HarperCollins.

Collins, J.C. and Porras, J.I. (1997). *Built to Last: Successful Habits of Visionary Companies*. New York: HarperCollins.

Drury, C. and Gramopadhye, A. (1991). Speed and accuracy in aircraft inspection. Position paper for FAA biomedical and behavioral sciences division. Washington, DC: Office of Aviation Medicine.

Fleming, M. and Lardner, R. (1999). Safety culture – the way forward. *The Chemical Engineer*, 16–18.

Freiberg, K. and Freiberg, J. (1996). *Nuts! Southwest Airlines' Crazy Recipe for Business and Personal Success*. Austin: Bard Press.

GAIN (2004). A roadmap to a just culture: enhancing the safety environment. A report by the global aviation information network working group E. In: Flight Safety Digest, March 2005. Flight Safety Foundation. Retrieved from <http://www.flightsafety.org> on November 14, 2005.

Gallagher, C., Underhill, E., and Rimmer, M. (2001). Occupational health and safety management systems: a review of their effectiveness in securing healthy and safe workplaces. A report for the National Occupational Health and Safety Commission, Sydney, Australia. Retrieved from <http://www.nohsc.gov.au/Pdf/OHSSolutions/ ohsms_review.pdf> on November 14, 2005.

Grubbe, D.L. (2003). Safety at DuPont. Presented at the US House Science Committee Hearing on NASA organizational issues, October 29, 2003. Retrieved from <http://www.house.gov/science/hearings/full03/oct29/ grubbe.htm> on October 12, 2005.

Hackworth, C., Cruz, C., Jack, D., Goldman, S., and King, S. (2004). Employee attitudes within the air traffic organization. Final report. Washington, DC: Federal Aviation Administration, Department of Transportation.

Harper, M. and Helmreich, R. (2003). Creating and maintaining a reporting culture. In: R. Jensen (ed.) *Proceedings of the Twelfth International Symposium on Aviation Psychology* (pp. 496–501), April 14–17, Dayton: The Ohio State University.

Helmreich, R.L. (1999). Building safety on the three cultures of aviation. In: *Proceedings of the IATA Human Factors Seminar* (pp. 39–43). Bangkok, August 12, 1998.

Herzberg, F. (1966). *Work and the Nature of Man*. Cleveland and New York: The Word Publishing Company.

Kaplan, R. and Norton, D. (1998). The balanced scorecard – measures that drive performance. In: *Harvard Business Review on Measuring Corporate Performance* (pp. 123–146). Boston: Harvard Business School Press.

Knowles, R.N. (2002). *The Leadership Dance: Pathways to Extraordinary Organizational Effectiveness*. Niagara Falls: The Center for Self-Organizing Leadership.

La Porte, T. and Consolini, P. (1991). Working in practice but not in theory: theoretical challenges of high-reliability organizations. *Journal of Public Administration Research Theory*, 1, 19–47.

Marx, D. (1997). Moving toward 100 per cent error reporting in maintenance. In: *Proceedings of the Eleventh International Symposium on Human Factors in Aircraft Maintenance and Inspection*. Washington, DC: Federal Aviation Administration.

Marx, D. (2001). Patient safety and the 'just culture': a primer for health care executives. Retrieved from <http://www.mers-tm.net/support/Marx_Primer.pdf> on October 11, 2004.

Patankar, M.S. (2002). Causal-comparative analysis of self-reported and FAA rule violation datasets among aircraft mechanics. *International Journal of Applied Aviation Studies*, 2(2), 87–100. Okalahoma City: FAA Academy.

Patankar, M.S., Bigda-Peyton, T., Sabin, E., Brown, J., and Kelly, T. (2005). A comparative review of safety cultures. Report prepared for the Federal Aviation Administration. Available from <http://hf.faa.gov>.

Patankar, M.S., Brown, J.P., and Treadwell, M. (2005). *Safety Ethics: Cases from Aviation, Medicine, and Environmental and Occupational Health*. Aldershot: Ashgate.

Patankar, M.S. and Driscoll, D. (2004). Preliminary analysis of aviation safety action programs in aviation maintenance. In: M. Patankar (ed.) *Proceedings of the First Safety Across High-consequence Industries Conference* (pp. 97–102) [CD-ROM], St. Louis.

Patankar, M.S. and Gomez, M. (2005). *Proceedings of the maintenance ASAP infoshare working group meeting*. June 27 and 28, 2005. St. Louis: Saint Louis University.

Patankar, M.S. and Sabin, E. (2008). Safety culture transformation in technical operations of the air traffic organization: project report and recommendations. Report prepared for the Federal Aviation Administration. Available from <http://hf.faa.gov>.

Patankar, M.S. and Taylor, J. (1999). Corporate aviation on the leading edge: systemic implementation of macro-human factors in aviation maintenance. In: *Proceedings of the SAE Airframe/Engine Maintenance and Repair Conference* [SAE Technical Paper Number 1999-01-1596]. Vancouver.

Patankar, M.S. and Taylor, J.C. (2004). *Risk Management and Error Reduction in Aviation Maintenance*. Aldershot: Ashgate.

Patankar, M.S., Taylor, J. and Goglia, J. (2002). Individual professionalism and mutual trust are key to minimizing the probability of maintenance errors. In: *Proceedings of the Aviation Safety and Security Symposium*, April 17 and 18, Washington.

Pierce, G. (2005). Building the foundation for a culture of safety. Focus on patient safety: a newsletter from the National Patient Safety Foundation, 8(2), 1–3.

Reason, J. (1997). *Managing the Risk of Organizational Accidents*. Aldershot: Ashgate.

Roberts, K.H. (ed.) (1993). *New Challenges to Organizations: High Reliability Understanding Organizations.* New York: Macmillan.

Roughton, J.E. and Mercurio, J.J. (2002). *Developing an Effective Safety Culture: A Leadership Approach.* Boston: Butterworth-Heinemann.

Schulman, P. (1993). The analysis of high reliability organizations: a comparative framework. In: K.H. Roberts (ed.), *New Challenges to Organizations: High Reliability Understanding Organizations.* New York: Macmillan.

Senge, P.M. (1990). *The Fifth Discipline: The Art and Practice of the Learning Organization.* New York: Doubleday/Currency.

Sloat, K. (1996). Why safety programs fail ... and what to do about it. *Occupational Hazards*, 58(5), 65–70.

Strebel, P. (1998). Why do employees resist change? In: *Harvard Business Review on Change* (pp. 139–158). Boston: Harvard Business School Press.

Taylor, J.C. (1995). Effects of communication and participation in aviation maintenance. In: *Proceedings of the Eighth International Symposium on Aviation Psychology* (pp. 472–477). Columbus: The Ohio State University.

Taylor, J.C. (2000). Reliability and validity of the maintenance resources management/technical operations questionnaire. *International Journal of Industrial Ergonomics*, 26, 217–230.

Taylor, J.C. (2004). Prototype training materials for acceptance criteria of maintenance ASAP events occurring within social context. Moffett Field: QSS/NASA Ames Research Center. Retrieved from <http://www.faa.gov/safety/programs_initiatives/aircraft_aviation/asap/ reports_presentations/media/MaintenanceASAP.pdf> on November 14, 2005.

Taylor, J.C. and Christensen, T. (1998). *Airline Maintenance Resource Management: Improving Communication.* Warrendale: Society of Automotive Engineers.

Taylor, J.C. and Fenton, D.F. (1993). *Performance by Design: Sociotechnical Systems in North America.* Englewood Cliffs: Prentice Hall.

Taylor, J.C. and Patankar, M.S. (2001). Four generations of MRM: evolution of human error management programs in the United States. *Journal of Air Transportation World Wide*, 6(2), 3–32.

VanDrie, K.D. (2005). Risk and resource management: an analytical approach. In: Patankar, M. and Ma, J. (eds) *Proceedings of the Second Safety Across High-consequence Industries Conference*, Volume 1, St. Louis: Parks College of Engineering, Aviation and Technology.

Westrum, R. (1993). Cultures with requisite imagination. In: J. Wise and D. Hopkin (eds), *Verification and Validation: Human Factors Aspects* (pp. 401–416). New York: Springer.

Westrum, R. (1995). Organisational dynamics and safety. In: N. McDonald, N. Johnston and R. Fuller (eds), *Applications of Psychology to the Aviation System* (Vol. 1) (pp. 75–80). Brooksfield: Ashgate.

Wheatley. M.J. (1999). *Leadership and the New Science: Discovering Order in a Chaotic World* (second edition). San Francisco: Berrett-Koehler Publishers.

Zahlis, D. and Hansen, L. (2005). Beware [of] the disconnect: overcoming the conflict between measures and results. *Professional Safety*, 18–24. American Society of Safety Engineers.

Chapter 5
Safety Strategies

Introduction

A strategy is an overarching plan toward a specific goal. The *Safety Strategies* section of the Safety Culture Pyramid comprises leadership strategies; organizational mission, values, structures and goals; processes, practices, and norms; and history, legends, and heroes. These are some of the most tangible aspects of safety culture. They can be discovered through a variety of means: field observations, artifact analysis, interviews and focus group discussions, and dialogue. A specific strategic plan toward achieving specific safety goals may or may not be available or published; however, leaders' strategies, organizational structures, routine processes and practices, legends and heroes are always present, and they collectively describe the extant safety strategies and their goals (whether intentional or not).

At the macro level, safety strategies could be broadly classified as value-based or compliance-based. Value-based strategies are derived from factors internal to the organization – a company or a hospital decides to change because it is the right thing to do. Compliance-based strategies are derived from external pressures, regardless of whether they are due to regulatory mandates or business mandates. In most cases, however, safety strategies are a result of a combination of the two – there's sufficient level of readiness for change in an organization and the external pressures serve as catalysts, accelerating the adoption of the changes. Compliance-based strategies tend to be reactive and focused on the short-term goals; value-based strategies tend to be proactive and focused on the long-term goals.

As discussed in previous chapters, the state of the safety culture may be expressed along two scales: accountability and learning. Data retrieved from field observations, artifact analysis, interviews and focus group discussions, and dialogue could be presented in the form of a 2×2 matrix wherein five elements of a safety strategy are mapped across a maturity rating of one to four. For example, in one organization, the safety performance may not be measured at all (rating equals one); whereas, for an organization on the other end of the scale, the performance may be ranked above average in the industry sector (rating equals four). Conceptually, the safety strategies maturity index corresponds to the two scales as follows: on the accountability scale, one equals secretive culture, two equals blame culture, three equals reporting culture and four equals just culture; and on the learning scale, one equals failure to learn, two equals isolated learning, three equals continuous learning, and four equals transformative learning. Table 5.1 provides examples of safety strategies mapped across the rating scales.

As an organization's safety strategies mature, they move from a compliance-based approach to a value-based approach.

This chapter continues to explore the Comair accident and the heparin dosage errors to illustrate how actual safety strategies can be brought to light. Examples of a number of processes and programmes used to improve safety culture in aviation as well as healthcare are presented throughout this chapter. Toward the end, two examples are presented: one is from aviation and the other from healthcare. Although there are differences in these two strategies, their ultimate goal is to transform the safety culture.

Table 5.1   Safety strategies maturity index

	Safety Strategies Maturity Index			
	1	2	3	4
Cultural states				
Accountability scale	Secretive	Blame	Reporting	Just
Organizational learning scale	Failure to learn	Isolated learning	Continuous learning	Transformative learning
Safety performance	Undocumented/ poor	Below average	Average	Above average
Safety goals	Hide/ignore the problem	Solve the specific problem	Solve the cluster of problems	Build a system of addressing latent factors that may lead to problems
Safety tactics	Cover up	Terminate the employee who was closest to the problem	Address systemic issues that led to the problem	Build a 'system of systems' to continuously scan for signs of deterioration in safety
Safety resources	None allocated	Emergency or one-time allocation	Minimal budget to solve specific problems	On-going investments in safety initiatives
Safety assessment	None	Episodic	Regular, but focussed on specific problems	Regular, systemic, and forward-looking
Safety communication	None	Need-to-know basis	Regular, but singular medium	Intense, multiple media

Leadership Strategies

Safety leadership in a high-consequence industry is about creating an environment in which safety will be practiced as an enduring value. Going back to the Safety Culture Pyramid, the values and unquestioned assumptions at the base of the pyramid should be lived by the leaders. In fact, the Institute of Nuclear Power Operations (INPO, 2004) defines safety culture as, 'an organization's values and behaviours – modeled by its leaders and internalized by its members – that serve to make nuclear safety the overriding priority'. If safety is not an explicit value or it is not specifically listed in the mission statement of the organization, it needs to be. So, the first task of a leader should be to get safety listed as a key organizational value and/or get it incorporated in the organizational mission statement. Next, there should be appropriate organizational structures, procedures, and policies that operationalize and institutionalize safety as a corporate value. For example, there should be a director of safety who reports directly to the President/CEO and is empowered to implement safety-related improvements across the organization; there should be a non-punitive reporting system that enables the employees to report errors and hazards without any fear of reprisal; there should be appropriate recognition systems that applaud employee and management efforts to improve safety; there should be equal consideration to safety issues as there is to production and technical competency issues in the annual evaluation of employees and management; there should be an annual safety climate survey assessing the safety attitudes and opinions of all the employees; and if an undesirable event occurs, emphasis should be on finding the systemic contributors and implementing a systemic solution. If employees have performance issues, they should be addressed outside of safety violations (don't use safety violation data to penalize a poor performer).

Well, this sounds like a great plan if you are starting your own company and are able to hire all your employees, and are in a position to unilaterally influence the organization. Most of us are not so lucky! We are placed in a system that was designed by others, consisting of a variety of employee/management personalities, regulatory complications, operational challenges, financial challenges, and limited leverage. So, where do we start?

Opportunities to take a leadership role exist at all levels of an organization. Whether you are a frontline employee, a middle manager, or the President/CEO, you have opportunities to demonstrate your commitment toward safety. Often, these opportunities start as single incidents, but they offer large-scale learning and transformational potential. For example, the families of patients who have been victims of medical error have a choice: (a) file a lawsuit against the hospital or the physician and seek financial compensation for their loss or (b) turn the tragic situation into a transformative experience for all. It is certainly not easy to put aside all the feelings of anger and mistrust against the hospital/physician, but there are some truly remarkable examples of very courageous families who have overcome their anger and turned their tragedy into success for the thousands of patients and healthcare personnel.

What should leaders do? The Institute of Nuclear Power Operators (INPO, 2004) lists the following eight principles of a strong safety culture:

- Everyone is personally responsible for nuclear safety.
- Leaders demonstrate commitment to safety.
- Trust permeates the organization.
- Decision-making reflects safety first.
- Nuclear technology is recognized as special and unique.
- A questioning attitude is cultivated.
- Organizational learning is embraced.
- Nuclear safety undergoes constant examination.

Executives and senior managers are expected to lead by example. Across the list of behavioural attributes specifically listed for leaders, the following themes are evident:

- Hands-on leadership: close monitoring of daily activities, coaching of people, and zero tolerance for deviations from published procedures or trained behavioural expectations.
- Open communication with employees: maintain clear and open communication lines so that nobody is afraid to speak up or express a different opinion. In fact, the managers are expected to encourage diverse perspectives and make the most conservative decision.
- Sensitivity toward the appearance of production goals overriding safety goals: managers are expected to retain safety as their top priority, regardless of the production pressures.
- Safety is an integral part of personnel selection and evaluation processes: managers are expected to give safety issues just as much importance in annual evaluations as they give to production targets.

In the context of safety culture transformation, leaders must: (a) articulate a clear vision with measurable goals along with the rationale to seek these goals, (b) engage the subordinate managers and employees in developing appropriate strategies and tactics to achieve the vision/goals, (c) align the reward/recognition systems to ensure that appropriate priority is placed on achieving the safety goals, (d) provide the necessary resources, and (e) foster open communication across the vertical and horizontal lines within the organization.

Organizational Mission, Values, Structures, and Goals

Organizational mission statements, list of values, organizational structure, and a list of strategic goals serve as valuable artefacts of safety culture. Some of these artefacts may change over time, but such changes also tell a story about the organization's evolution. Further, operational goals and performance expectations from individual through organizational levels could be tied to each other. Ideally, every employee should be able to link his/her daily actions to one of the organizational goals.

Let us take the examples of Comair 5191 and heparin administration errors, presented in Chapter 3 and delve deeper into the industry-wide trends, and then further into the safety strategies undertaken by various stakeholders (airlines/ hospitals, industry-wide partner organizations, and certifying agencies).

Corporate Perspective

A review of how different organizations respond to crises provides an interesting first impression of the prevailing cultural state in that organization. Such a response, however, needs to be examined from the perspective of that particular organization's mission, values, historical context, and role/tenure of its top leadership. In this section, we present the corporate perspective for Comair, Cedars-Sinai, Methodist Hospital, and CHRISTUS-Spohn Hospital.

Comair In 1977, Comair was founded by a father-son pair, Raymond and David Mueller. They launched their airline with only three airplanes and service to four cities. Since then, the company grew and started developing a business partnership with Delta Airlines. The Comair-Delta relationship allowed Comair to grow as a regional connector of domestic and international services. By 1997, the company had grown so large that it transported 5.5M passengers. In 1999, Comair won the Best Managed Regional Airline Award and Delta won the Best Managed Major Airline Award from *Aviation Week and Space Technology*. In 2000, Delta acquired Comair for $2.3B. In September 2005, Delta filed for Chapter 11 bankruptcy protection and took Comair with it. This resulted in over 1,000 job reductions and across the board reduction in compensation packages at Comair.

Comair lists the following corporate values:

- Safety is our first priority.
- Innovation drives our success.
- Caring sets us apart.
- Speed keeps us on the forefront.
- Simplicity makes better business.
- Service guides our culture.
- Teamwork ties it all together.

Comair's mission statement was not readily available, but since it is a wholly-owned subsidiary of Delta Airlines, Delta's mission and values statements are presented:

> The mission of Delta Air Lines is to build on its traditions and always meet customers' expectations while taking service to ever higher levels of excellence. The company stands for safe and reliable air transportation, distinctive customer service, and hospitality from the heart.

While the mission statement itself is typical of any organization in the service industry, the values listed in the second sentence clarify that 'safe and reliable air transportation' is one of the three key values of the organization. This value is consistent with Comair's values.

Next, a look at the tenure of Comair's president reveals that there has been a new president every year from 2005 through 2008:[1] Fred Buttell (2005); Don Bornhorst (2006); John Selvaggio (2007); and John Bendoraitis (2008). The point we wish to make here is that relying on the President or the Chief Executive Officer to lead the safety campaign is not always practical. However, if safety is an enacted organizational value, then it won't matter who the President/CEO is and how long he/she stays in the position. So, how did Comair respond to the August 2006 tragedy? First, the company released a statement expressing deep condolences to the families of passengers who died in the accident. Second, it accepted responsibility for the pilots' errors. Then, it filed legal suits against the Blue Grass Airport for improper management of the taxiway construction project and against the FAA for allowing only one air traffic controller to man the tower at the time of the accident. The alleged intent of the lawsuits was not to recover any financial compensation, but to establish the fact that multiple factors had contributed to the accident, and pilot error was not the sole factor, laying the groundwork for a percentage of the damages to be recovered from the FAA. Next, the company tried to stop confidential safety reports being released to the plaintiffs' attorneys. It was important to guard these reports because its pilots had submitted these reports voluntarily and with the expectation that the reports would: (a) contribute toward system-wide improvements, (b) not be used against the pilots or any punitive actions, and (c) not be made public. Ultimately, when Comair was compelled to release the pertinent Aviation Safety Action Program (ASAP) reports, the integrity of the ASAP programme was in question. In October 2008, Comair terminated its ASAP programme. Later, in May 2009, the programme was reinstated, reaffirming Comair's commitment toward safety.

Cedars-Sinai Hospital, Methodist Hospital, and CHRISTUS-Spohn Hospital Cedars-Sinai is founded on the Judaic tradition of care and giving

---

1 <http://www.comair.com/comair/cdc.portal?_nfpb=trueand_pageLabel=cdc_pg_0008>.

through healing. It is the largest non-profit hospital in the western United States and is widely regarded as Southern California's 'gold standard' in healthcare. In 2009, *US News and World Report* ranked Cedars-Sinai among the nation's best hospitals; eleven of its specializations were ranked. Further, Cedars-Sinai is among the Los Angeles area's top employers – 10,000 employees – winning several consumer awards. Thus, in many ways, it is regarded as a nationally and internationally renowned hospital.

Cedars-Sinai Health System[2] is committed to:

- Leadership and excellence in delivering quality healthcare services.
- Expanding the horizons of medical knowledge through biomedical research.
- Educating and training physicians and other healthcare professionals.
- Striving to improve the health status of our community.

Throughout its statement of values and vision, there appear to be three key themes: top quality clinical service and compassionate patient care; basic and clinical research that benefits the patients; and top quality medical, health sciences, and community health education.

So, how did Cedars-Sinai respond to the heparin error reported in Chapter 3? First, they accepted responsibility for the error. In a statement[3] made by Dr. Michael L. Langberg, M.D., Chief Medical Officer, Cedars-Sinai Medical Center, he acknowledged that the medical error was preventable. Second, Dr. Langberg apologized to the families who were affected by this error and promised to work with them on their concerns and questions. Third, he acknowledged the seriousness of the situation and committed to a comprehensive investigation. Finally, he said that they would 'take all necessary steps to ensure that this never happens here again'. This final statement is very important because it sets a safety performance target for the hospital, making it clear to the internal and external stakeholders that the hospital has committed to zero preventable error – at least in the case of heparin administration.

The Methodist Hospital in Indianapolis is a part of Clarian Health and is closely tied with Indiana University's School of Medicine. Clarian Health's mission[4] is 'to improve the health of our patients and community through innovation and excellence in care, education, research and service'. Their values are as follows:

- Total patient care, including mind, body and spirit.
- Excellence in education for healthcare providers.
- The highest quality of care and respect for life.

---

2   <http://www.csmc.edu/31.html>.
3   <www.cedars-sinai.edu/pdf/Statement-11-20-07-56336.pdf>.
4   <http://www.clarian.org/portal/Clarian/about-clarian-health?ContentID=/about/living-our-mission.xml>.

- Charity, equality and justice in healthcare.
- Leadership in health promotion and wellness.
- Research excellence.
- An internal community of mutual trust and respect.

As with Cedars-Sinai, Clarian Health also values quality patient care, and patient safety is generally embedded under healthcare quality. However, in the case of Clarian, there is a visible emphasis on the following patient safety initiatives:[5]

- Computerized physician order entry.
- Computerized medical stocking and distribution.
- Re-education and training for nurses and clinical staff for the administration of high-alert drugs.
- Continued compliance with Joint Commission National Patient Safety goals, which include: improving the accuracy of patient identification; improving the effectiveness of communication among caregivers; improving the safety of using medications; and encouraging the active involvement of patients and their families in patient care as a patient safety strategy.

So, how did Clarian/Methodist Hospital respond to the heparin errors that resulted in the death of three babies? Two babies died on September 16 and the third baby died on September 19, 2006. On September 20, Methodist Hospital released a document called 'Heparin Error Backgrounder'[6] explaining the incident, the specific steps taken by hospital management to respond to the crisis as well as to prevent such an error in the future. This statement reinforced Clarian's commitment to open communication by saying, 'our culture expects forthrightness when mistakes are made so that we can learn from those mistakes and improve our systems and offer higher levels of patient safety'. Further, the statement clarified that this was a case of 'institutional error' and therefore, they expect the staff to return to work, after a short personal leave. Clearly, there would be expectation of more stringent compliance with existing safeguards and adherence to the new procedural requirements when handling high risk medications.

Clarian President and CEO Daniel F. Evans, Jr., said in a reaffirming comment, 'Clarian is devoted to patient safety and has many programs in place to enhance our quality of care. Our goal is to be error-free, and we will continue to dedicate both financial and staff resources to achieve this goal.' Again, similar to Cedars-Sinai, the organizational commitment to patient safety is strong and quantifiable.

---

5   <http://www.clarian.org/portal/Clarian/about-clarian-health?ContentID=/about/patient-safety/index.xml>.

6   Heparin Error Backgrounder drafted September 20, 2006. Available at <http://www.postdoc.medicine.iu.edu/>.

CHRISTUS-Spohn Hospital's mission[7] is to 'extend the healing ministry of Jesus Christ', and its values are expressed in terms of a guarantee statement[7]. 'We guarantee to those we are privileged to serve:

- Courteous, prompt and compassionate care.
- Concern for your special needs and privacy.
- Open, honest communications about your treatment.'

In the case of Spohn, the heparin error was different from that at Cedars-Sinai or the Methodist Hospital; here the error occurred when heparin was mixed with other solutions, including saline, in their pharmacy. The overdose was a result of too much of the paediatric version of heparin, not a mix-up between adult and paediatric concentrations (Chirinos, 2008). So, how did Spohn respond to erroneous administration of heparin to 17 babies? Chirinos reported the following comment from Spohn's CEO, Bruce Holstein, in response to the heparin errors: 'We had a stand down in the pharmacy department related to compounding and reinforced our policy regarding compounding. We reiterated our double-verification process.' In the double verification process, the pharmacy technician prepares the medication and the pharmacist cross-checks it, notes Chirinos.

Industry Perspective

In this section, we provide an overview of how the two industries, aviation and healthcare, responded to the two errors. The point we are trying to make here is that there are certain aspects of safety culture that transcend organizational boundaries and take root across the profession/industry. Consequently, when a particular professional practice is challenged or an error is publicized, there is likely to be an industry-wide response. In the case of runway incursion errors, you will see that it is a top-priority issue for the FAA, major as well as regional airlines, general aviation, and multiple labour unions such as the Air Line Pilots Association. Together, they are all taking specific steps toward reducing runway incursions. Similarly, in the case of heparin dosage errors, you will note that there is leadership and concerted efforts from industry groups like the Institute for Healthcare Improvement, the Agency for Healthcare Research and Quality, the National Patient Safety Foundation, as well as several hospitals and some medical equipment manufacturers. The Federal Drug Administration and the state-level Departments of Public Health are also taking an active role in investigating medical errors. Of course, penalty-based systems such as litigation and insurance premiums also drive reform in both industries.

---

7   <http://www.christusspohn.org/service_guarantee/guarantee.html>.

Aviation Industry

In the aviation industry, the Federal Aviation Administration (FAA) plays a critical role in assuring and improving safety. Typically, the FAA responds to safety issues via three types of strategic initiatives: training, research, and regulation. In the case of runway incursions, the goal of the training initiative is to raise the overall awareness of the issue and encourage all parties (pilots, mechanics, air traffic controllers, ground handlers, etc.) to recognize potential for runway incursions and take appropriate evasive actions. The goal of research initiatives is to continuously develop better means to prevent runway incursions. Such means include technical interventions like the Runway Status Lights System or improved training. Finally, the FAA may engage in rulemaking based on research and training data as well as widespread industry participation in crafting the final language of the rule.

It is not publicly known what specific actions Comair took to address the pilot complacency and lack of cockpit discipline. However, on a broader scale, runway incursion is a serious industry-wide problem. The following measures have been implemented to reduce runway incursions:

- The FAA issued the following guidance materials:
  - *Advisory Circular (AC) 120-74A, Flight Crew Procedures During Taxi Operations*, September 26, 2003.
  - *Safety Alert for Operators (SAFO) 06013, Flight Crew Techniques and Procedures That Enhance Pre-takeoff and Takeoff Safety*, September 1, 2006.
  - *SAFO 07003, Confirming the Takeoff Runway*, April 16, 2007.
  - *Information for Operators (InFO) 07009, Runway Lights Required for Night Takeoffs in Part 121 Operation*s.
  - InFO 07018, Taxi Clearances: Know the Rules, Understand Your Clearance.
  - On August 15, 2007, the FAA held a 'Call to Action' to focus on short-term and long-term measures to further improve runway safety. The participants included representation from the three key groups: airports, air carriers, and air traffic control. Each group was asked to present specific recommendations to the FAA.
- The Air Transport Association (ATA) and the Regional Airline Association (RAA) have assembled a list of initiatives that they would like to implement.
- Major labour groups such as the Air Line Pilots Association (ALPA), the Allied Pilots Association, and the Southwest Pilots Association all agreed to endorse and support the air carriers' initiatives, which are based on ATA and RAA initiatives.
- ALPA developed an online training programme for pilots. It is sponsored by the FAA's Office of Runway Safety and Operational Services.
- The FAA completed a survey on October 15, 2007. This survey determined the extent to which part 121 air carriers have implemented the 'Call to Action' initiatives. The survey showed that over 92 per cent of the reporting

air carriers were in full voluntary compliance, which represents more than 96 per cent of the aircraft involved.
- In February 2009, President Obama signed the American Recovery and Reinvestment Act (ARRA or 'stimulus' funding) into law. Accordingly, the FAA was allocated $1.3B for a variety of projects that are critical to its mission, but would also be consistent with the purpose of ARRA – to stimulate the economy by creating/preserving jobs and investing in transportation, environmental protection, and other infrastructure needs for the long-term recovery of the American economy. The FAA invested $20M in airport lighting, navigation, and landing equipment. The Runway Status Lights System is designed to work with a ground-based radar system, known as Airport Surface Detection Equipment Model X (ASDE-X), to alert the pilots and ground vehicle operators with red lights along the runway when the runway is being used by another vehicle (aircraft or ground vehicle). Three airports have already implemented this system and nineteen other airports will do so by 2012.[8]

One very important unintentional side-effect of the Comair 5191 accident was the request for release of Aviation Safety Action Program (ASAP) reports filed by Comair pilots. Until then, the industry had operated with the understanding that such reports were protected from disclosure under 14 C.F.R. § 193. The plaintiffs sent a deposition notice to Comair's person most knowledgeable about the ASAP programme and requested release of ASAP reports pertaining to Comair pilots involved in wrong runway departures. This notice created an avalanche of response from the aviation industry. Airlines, pilot unions, and even the FAA, were all concerned that disclosure of information submitted via the ASAP programme would endanger the programme and could cause serious harm to the safety culture in aviation. After almost two years of legal battle, Magistrate Judge James B. Todd's order to release the pertinent ASAP reports was upheld by United States Senior Judge Karl S. Forestor of the Eastern District of Kentucky. The following points are made in the Opinion and Order:

- There was no statutory or regulatory privilege protecting the ASAP reports from discovery in litigation.
- The United States Congress did not create a statutory privilege specifically for ASAP or other voluntary safety reports.
- If disclosure of ASAP reports to litigants under a confidentiality order presents a serious danger to aviation safety, Comair and the *amici* should implore the FAA or Congress to change the regulations or statute to preclude disclosure to litigants rather than authorizing disclosure pursuant to a court order as the regulations do now.

---

8   FY2009 FAA Performance and Accountability Report. Available at <http://www.faa.gov/about/plans_reports/media/2009_par.pdf>.

- 14 C.F.R. §193 merely restricts disclosure by the FAA; it does not create a privilege for the airline to prevent disclosure in subsequent litigation.
- The protection from public disclosure, provided by 14 C.F.R. §193 permits disclosure 'to correct a condition that compromises safety or security, if that condition continues uncorrected', 'to carry out a criminal investigation or prosecution', and when 'ordered to do so by a court of competent jurisdiction'.

Ultimately, the reports were disclosed and out of the multiple plaintiffs that received compensation in the Comair case, it is known that one party received $7.1M; settlements with other parties are confidential (Steitzer, 2009).

Healthcare Industry

In the American healthcare industry, several national organizations are working together and with their respective constituencies to individually as well as collectively improve patient safety. Following is a list of key organizations influencing the national patient safety agenda:

- The Institute of Medicine.
- The Agency for Healthcare Research and Quality.
- The Joint Commission on Accreditation of Healthcare Organizations.
- The Institute for Healthcare Improvement.
- The National Patient Safety Foundation.
- The Institute for Safe Medication Practices.

The Institute of Medicine[9] (IOM), established in 1970, is an independent non-profit organization affiliated with the National Academy of Sciences. The IOM 'serves as adviser to the nation to improve health'. As such, IOM reports are highly regarded across the nation, not just in the healthcare community. The seminal report, *To Err is Human: Building a Safer Health System*,[10] released in November 1999, raised public awareness about the gravity of medical errors and launched a flurry of activities across the various healthcare groups. The Agency for Healthcare Research and Quality[11] (AHRQ), a part of the Department of Health and Human Services, was originally established as the Agency for Healthcare Policy and Research in 1989. The goal of AHRQ is to conduct and sponsor research that benefits the healthcare community as well as policymakers. The Joint Commission's lineage[12] dates back to 1910 when the idea of standardization across hospitals was initially proposed by Dr. Ernest Codman; the first set of hospitals

---

9 <http://www.iom.edu/>.
10 <http://books.nap.edu/openbook.php?record_id=9728>.
11 <http://www.ahrq.gov/>.
12 <http://www.jointcommission.org/AboutUs/joint_commission_history.htm>.

were certified in 1918 by the American College of Surgeons; the Joint Commission was established in 1951. As an accrediting agency, the Joint Commission plays a pivotal role in influencing the performance standards and data collection processes used in hospitals across the nation. The Institute for Healthcare Improvement[13] (IHI), founded in 1991, is an independent not-for-profit organization with the mission 'to lead the improvement of healthcare throughout the world'. IHI's main contribution has been the development of toolkits and educational materials that accelerate as well as standardize the adoption of proven practices across hospitals. The National Patient Safety Foundation[14] is a non-profit organization established in 1997; its goal is 'to improve the safety of patients'.

With respect to the heparin administration error, the broader issue is medication error or adverse drug event, because heparin is not the only drug that is involved in administration errors and the root causes of such errors are very similar. Therefore, it is useful to review the healthcare industry's overall position on adverse drug events.

- Prevention of medication errors has been a National Patient Safety Goal of the Joint Commission each year since at least 2003.[15] These goals automatically become requirements for each of the accredited hospitals. Therefore, each accredited hospital has to produce evidence of how they are complying with the National Patient Safety Goals.
- AHRQ has published research findings[16] on adverse drug events – the causes, the scope of the problem, the financial impact on the hospitals, and suggested prevention strategies.
- IHI has created a number of tools, toolkits, and reference materials as resources[17] for those interested in reducing adverse drug events.
- NPSF holds the annual Patient Safety Congress to provide a forum for practitioners and researchers to share best practices and latest knowledge about patient safety programmes.
- The National Patient Safety and Quality Act of 2005[18] allows for the creation of Patient Safety Organizations, which may collect voluntarily submitted error reports. The Act provides strong protection of voluntarily submitted patient safety reports from subpoena, order, discovery, or Freedom of Information Act (FOIA) request with respect to Federal, State, local, civil or criminal cases. This is a much stronger protection than that afforded by 14C.F.R. §193 for aviation safety reports.

---

13 <http://www.ihi.org/ihi/about>.
14 <http://www.npsf.org/>.
15 <http://www.jointcommission.org/PatientSafety/NationalPatientSafetyGoals/03_npsgs.htm>.
16 <http://www.ahrq.gov/qual/aderia/aderia.htm>.
17 <http://www.ihi.org/IHI/Topics/PatientSafety/MedicationSystems/Tools/>.
18 <http://frwebgate.access.gpo.gov/cgi-bin/getdoc.cgi?dbname=109_cong_public_lawsanddocid=f:publ041.109>

- The Institute for Safe Medication Practices[19] is a nonprofit organization dedicated to 'medication error prevention and safe medication use'. It is a federally certified Patient Safety Organization and runs a voluntary, national medication error reporting programme under the protection of the National Patient Safety and Quality Act of 2005. The Federal Drug Administration monitors the reports[20] submitted through this programme.

Specifically, in the case of heparin errors at Cedars-Sinai, the California Department of Public Health issued citations against Cedars-Sinai for the following:

- Multiple failures by the facility to adhere to established policies and procedures for safe medication use;
- Caused, or were likely to cause, serious injury or death to the patients who received the wrong medication; and
- Created a risk of harm for all hospital patients.

Although the State law allows for a maximum fine of $25,000 *per* violation, Cedars-Sinai was fined a *total* of $25,000. A 2006 study reported that the average medical malpractice claim with medical errors was awarded $521,560[21] per claim. Further, hospitals and healthcare organizations may face vicarious liability claims for failure to address known systemic problems, particularly when state agencies have fined the facility for non-compliance.

Clearly, as the awareness of preventable medical errors grows, coupled with the potential fines by state agencies and vicarious claims by plaintiffs, both industry-wide as well as organization-level interventions will be tracked more seriously.

Processes, Practices, and Norms

Going back to the Heinrich ratio discussed in the previous chapters, it is generally believed that for every fatal accident, there are about 39 non-fatal, but significant incidents and about 300 unreported incidents. In order to address the systemic latent failures that create error-producing conditions, both aviation and healthcare are striving to reduce the number of significant incidents (at the 39 level of the Heinrich ratio) as well as increase the number of reports of errors (at the 300 level of the Heinrich ratio) so that error trajectories can be intercepted with effective safety nets and harm can be prevented. Typically, accident/incident reports or error reports result in changes in previously established processes, practices or norms.

---

19 <http://www.ismp.org/about/default.asp>.
20 <http://www.fda.gov/Drugs/DrugSafety/MedicationErrors/default.htm>.
21 Claims, errors, and compensation payments in medical malpractice litigation. *New England Journal of Medicine*, May 11, 2006.

Processes are published protocols of how a particular task is to be performed. Practices are the actual execution of actions to perform the task; however, sometimes the practices are either different from the published procedures or published procedures may not exist for certain tasks. Norms are generally blatant violations of published procedures because of convenience, lack of resources, or a belief that the published procedure is ineffective, erroneous, or not needed. Norms develop over time because the published procedures are impractical to follow and the people develop their own way of accomplishing the mission and bypass the published process. Fatal accidents at the top of the Heinrich ratio are believed to be preventable if the precursors to those accidents are understood and mitigated in time. From a strategic perspective, the goal is to change the underlying structures, processes, practices, norms, and technologies to eliminate the error-producing conditions, thereby minimizing the probability of a catastrophic (fatal) failure.

Safety Performance Metrics

In 1997, The White House Commission on Aviation Safety and Security, chaired by Vice President Al Gore, recommended specific safety performance goals for the US airline industry: compared to the 1994–96 baseline data, to reduce the risk of fatal accidents by 80 per cent by 2007 and to reduce the total number of accidents per year. In response to this recommendation, the FAA launched the Safer Skies initiative, which resulted in a unique industry-government partnership in the form of three teams: (a) the Commercial Aviation Safety Team (CAST), (b) the General Aviation Joint Steering Committee, and (c) Partners in Cabin Safety. McVenes (2007) reported on the success of the CAST team in helping achieve a 76 per cent reduction in the risk of fatal accidents. This is a particularly remarkable achievement because in March 1997, the General Accounting Office had released a report casting doubts on the industry's ability to achieve this lofty goal. Highlights of the CAST team's approach are presented here to illustrate how accident/incident data could be used to develop priorities and performance targets, which would then lead to specific strategies, tactics, and measurable success.

The CAST team formed three working groups: the Joint Safety Analysis Teams (JSAT), the Joint Safety Implementation Teams (JSIT), and the Joint Implementation Monitoring-Data Analysis Teams (JIMDAT). Using a data-driven approach, the JSATs identified a chain of events leading to each category of accident, determined intervention strategies, and evaluated the effectiveness of the intervention strategies. Then, the JSITs evaluated the feasibility of the intervention strategies that JSATs had proposed and if feasible, developed plans to implement the strategies. Finally, the JIMDATs monitored the field implementation of the strategies, recommended modifications, if necessary, and reported progress back to the full CAST team. Airlines, pilots, unions, manufacturers, FAA, and Transport

Canada representatives participated in the CAST Team. Over the ten-year period, CAST recommended 65 interventions; 40 of which had been completed by May 31, 2007; and the remaining 25 were being implemented.

A review of all the CAST recommendations[22] (original 65 plus 'remaining risk' recommendations) reveals the following distribution (some are counted in multiple categories):

- Publications: 23.
- Training: 20.
- Technology Upgrades: 18.
- Procedural Changes:10.
- Programme Addition: 4.

It is not surprising that publications, training, and technology upgrades are the top three categories because clear and up-to-date publications will minimize misunderstanding among different crews, air traffic controllers, and mechanics. Also, new publications lead to additional/recurrent training. Training in itself is always a critical item for flight safety.

In spite of all the above changes and the associated claim of reducing the risk of fatal accidents by 76 per cent, we still don't have robust enough metrics that tightly couple such initiatives with periodic measurements. For example, it would be good to conduct surveys frequently enough to examine trends in specific behavioural patterns as well as attitudes and opinions. Similarly, regular analysis of Flight Operations Quality Assurance (FOQA), Aviation Safety Action Program (ASAP), and Line Operations Safety Audit (LOSA)[23] programmes, in conjunction with the above listed enhancements, would yield correlational data that could strengthen the argument regarding reduction in risk. Also, it would be easier to show evidence of whether or not the changes resulting from the enhancements were sustained.

Another challenge in aviation is that organizational safety performance goals are not tied to individual performance goals. For example, we have not yet seen any airline establish a goal of zero runway incursions, follow up with an analysis of precursors to runway incursions, provide appropriate training and guidance materials to the crew, and ultimately hold the crew accountable for managing the cockpits such that the precursors to runway incursions are eliminated. Also, there are no known examples of airlines tying financial incentives to safety performance goals.

---

22 <http://www.cast-safety.org/factsheets.cfm>.
23 FAA Voluntary Programs Branch: <http://www.faa.gov/about/office_org/headquarters_offices/avs/offices/afs/afs200/branches/afs230/>.

Training

Broadly, there are two types of safety training programmes: awareness programmes and behavioural programmes. The awareness programmes are aimed at raising the awareness of the participants about key safety issues. Typically, such programmes use a case-study approach to illustrate a particularly relevant accident and then discuss the causal factors and how these causal factors relate to the specific roles and responsibilities of the audience. While one might hope that awareness of causal factors may influence behaviour, there are no assurances. In fact, Patankar and Taylor (1999) discovered that awareness training programmes have a significant influence on the attitudes and opinions (safety climate) of the participants, but do not have an appreciable impact on their behaviours (safety performance). Further, the noted behavioural shift is limited to passive or self-focused changes (e.g. 'I am more aware of how fatigue influences my work' or 'I will try to listen more carefully during a shift-change briefing'). On the other hand, there were other programmes that focused directly on behavioural change. The Concept Alignment Process (Patankar and Taylor, 1999) is a structured communication protocol that may be used by all members of an aviation organization to communicate with each other within or across the traditional professional groups (pilots, mechanics, dispatchers, managers, etc.). Such direct focus on behavioural change caused changes to organizational policies and procedures, vendor contracts, and FAA documents.

The MRM programme has evolved over four generations (Taylor and Patankar, 2001), and is now moving toward the fifth generation (Patankar, 2009). Through the first four generations, the emphasis has been evolving from adaptation of crew resource management to maintenance environment, to focus on reduction of specific maintenance errors through open communication, to shift toward individual human performance factors in aviation maintenance, and toward integrating behavioural changes with increased awareness (Taylor and Patankar, 2001). The fifth generation programmes are beginning to integrate MRM training with maintenance ASAP programmes by using lessons learned from ASAP cases to improve maintenance systems as well as establish baseline standards of professional behaviour (Patankar, 2009).

Safety Management System in Aviation

The International Civil Aviation Organization (ICAO), a United Nations agency that establishes and enforces quality and safety standards across the world, has *required* all its 190 member States (meaning countries) to develop and implement Safety Management System (SMS) programmes. However, the FAA has not yet passed a rule to implement this international requirement within the United States. At the time of this writing an Advanced Notice of Proposed Rule Making (ANPRM) had been published and public comments received. One significant task in moving from the ANPRM stage to the Notice of Proposed Rule Making (NPRM) and then to a formal rule, is the need to demonstrate a business case for SMS.

SMS is a compliance-based safety strategy; nonetheless, it provides a comprehensive and standardized framework that the global aviation industry ('global' in the sense of multinational as well as inclusive of different industry sectors like flight operations, maintenance, and airports) could adopt. As in the nuclear industry, the SMS standard for aviation organization clearly delineates management commitment and responsibilities as follows:

> The standard specifies that top management is primarily responsible for safety management. Managements must plan, organize, direct, and control employees' activities and allocate resources to make safety controls effective. A key factor in both quality and safety management is top management's personal, material involvement in quality and safety activities. The standard also specifies that top management must further clearly delineate safety responsibilities throughout the organization (FAA, 2006).

The SMS programme consists of four pillars: 'Safety Policy, Safety Risk Management, Safety Assurance, and Safety Promotion' (FAA, 2006). Although this framework is presented in the aviation context, it could be applied to healthcare or any other industry.

Safety Policy

Under the SMS requirements, the SMS Safety Policy is expected to clearly articulate the management's commitment toward the following: implementation of SMS, continuous improvement of safety performance, management of safety risk, compliance with all the regulatory requirements, encouragement of employees to report safety issues without reprisal, establishment of clear standards of acceptable behaviour, guidance for management to set up safety objectives, periodic review of the set safety objectives, documentation of everything associated with SMS implementation, communication with all relevant parties, review of the SMS Safety Policy itself, and identification of the responsibilities of management and employees with respect to safety performance. Clearly, the development and implementation of such an all-encompassing policy is set to drive the organization toward a just culture, a strategic management of safety performance, and transformational learning.

Safety Risk Management (SRM)

The goal of SRM is to fully understand all the risks associated with specific actions and develop means to manage those risks to acceptable levels. Safety risk can be defined in terms of probability of failure and consequences of the failure. A

2×2 matrix of likelihood (probability) versus severity (consequences) is generally constructed to express the risk. A variety of technical, procedural, or human interventions/tactics could be applied to control the risk. The SRM approach is expected during all three key stages: design, operation, and maintenance. When products undergo a design change or when operating/maintenance procedures undergo a change, a thorough risk assessment is expected to ensure that new risks are not introduced during the change. Furthermore, effective feedback loops are expected to ensure that previously implemented safety risk control strategies are working. Therefore, if an error/hazard reporting system identifies a risk currently existing in the system (human, technical, procedural or organizational aspect), the SRM process will have to be invoked to either accept the risk at the extant level or manage it by implementing appropriate controls. Overall, the SRM process will make the safety risks clearly known to all parties involved, and a person or a group will need to formally accept the risks involved in the design, operation, and maintenance of a service or a product involved in air transportation.

Safety Assurance

Safety Assurance is analogous to Quality Assurance in the sense that just as quality assurance protocols are used to ensure that the quality of a product or service produced by an organization conform to pre-established standards and that appropriate means exist to measure and correct quality issues, there need to be safety assurance protocols to ensure that the safety of the product or service provided is maintained in accordance with the pre-established standards. The key components of safety assurance consist of the following: a system of hazard identification, ongoing risk assessment (under changing environmental conditions – business pressures, human performance issues, weather, hardware/equipment deterioration, etc.), and proactive management of risk.

Safety Promotion

The overall expectation under this pillar is assessment and improvement of safety culture, which the SMS Advisory Circular defines as 'the product of individual and group values, attitudes, competencies, and patterns of behaviour that determine the commitment to, and the style and proficiency of, the organization's management of safety' (FAA, 2006). Safety Climate surveys may be used to periodically measure the safety climate, compare the results with specific planned interventions or extraneous events that may have unintended consequences on safety culture. Marketing of success stories, lessons learned, and management's unwavering commitment toward a stronger safety culture are critical.

Global Implications

Besides improving safety at the local level, SMS programmes have the potential to enable consistency in the use of safety performance and safety climate metrics as well as assessment of various interventions across global benchmarks. For example, let's suppose that an airline decides to set a safety performance goal of 'zero unstabilized approaches' (windshear-related incidents would not be counted). This safety performance goal would be specified by the airline; appropriate baseline measurements would be made to quantify the scope of the risk (how often does it happen and what are the consequences); risk control tactics would be employed; and effectiveness of the control tactics would be measured. The airline's final approach data could be compared not only for all the flights within the airline, but if multi-airline data sharing is available (through a system like the Aviation Safety Information Analysis System[24]), data could be compared against other airlines (at the same airport or same type of aircraft) as well. The challenge in eliminating unstabilized approaches could involve crew training, technical interventions that provide early warning and guidance to correct the approach, change in air traffic control procedures, change in approach checklist, etc.

History, Legends, and Heroes

Every culture has heroes. These heroes are the epitome of values cherished by the culture at that time. Aviation has always been an industry of heroes and legends ... from the Wright Brothers to Captain Chesley Sullenberger. The aviation industry has always celebrated values such as courage, professional excellence, self-discipline, loyalty, and self-sacrifice. The Wright Brothers are celebrated not just because of their world-changing invention, but also because of their willingness to risk their own lives to test their conviction. Today, Captain Sullenberger is celebrated for his professionalism in the heroic landing of his US Airways airplane on the Hudson river after losing both engines. Similarly, there are hundreds of others who have flown the fastest, highest, longest, etc. In some ways, the efforts of hundreds of engineers, mechanics, ramp personnel, clerical staff, managers – essentially everyone except the pilot – are personified in the pilot's heroic achievement and celebrated. The Flight Safety Foundation[25] presents up to 14 awards each year for outstanding contributions – across all fields of aviation, including journalism and research – that enhance safety of flight.

---

24 <http://www.asias.faa.gov>.
25 <http://flightsafety.org/aviation-awards>.

Aviation Heroes

Among aviation safety champions, one of the most famous airline pilots is Captain Alfred C. Haynes. On July 19, 1989, Haynes was in command of the United Airlines flight 232, which lost power to the centre engine of a three-engine DC-10 and suffered complete hydraulic failure, crash-landing the airplane in Sioux City, Iowa, on its way from Denver to Chicago (NTSB, 1992). The possibility of a complete hydraulic failure, disabling the primary flight controls, was thought to be a mathematical improbability. However, when the engine suffered an uncontained failure, a turbine disk cut through hydraulic lines and spewed all the hydraulic fluid. For the aviation community, what makes Captain Haynes a hero is his ability to apply his crew resource management skills, his willingness to call upon a simulator instructor who was riding on the flight as a passenger, and work with the cockpit crew, cabin crew and air traffic controllers to land the airplane. As an airline pilot, Captain Haynes was a 'classic' pilot with military background and a long airline career. He connected very well with his contemporary pilots who had grown up in the military and were trained in the 'command and control' environment; however, his ability to work with his team both inside and outside the cockpit is what convinced the traditional pilot community of the merits of crew resource management.

John Goglia is a hero for mechanics, air crash investigators, fire fighters, and millions of airline employees. John started as a mechanic for United Airlines, went on to Allegheny Airways, which later became US Airways. While at US Airways, John worked on several safety committees and represented the International Association of Machinists and Aerospace Workers, a mechanics' labour union, in air crash investigations. In 1994, John was appointed to the National Transportation Safety Board by President Bill Clinton. John was the first aircraft mechanic to serve as a Board Member. John became the 'people's hero' because he genuinely cared about his colleagues and his extended family of airline employees; his ability to focus on the gruesome facts, void of political correctness, shocked most senior administrators and came as a welcome relief to most employees and passengers. John championed the cause of aircraft mechanics, brought to light the unintentional errors that mechanics can make and how the system is not set up to catch those errors, raised the awareness of maintenance resource management, and supported research in maintenance human factors.

On September 11, 2001, the unprecedented hijacker attacks tested the air traffic controllers and their managers. Hundreds of air traffic controllers, pilots, flight attendants, managers, and government officials became heroes because they acted swiftly and calmly to shut down the air space, declared a ground stop (prevent all aircraft from taking off), and brought down every flying aircraft (about 4,500!) safely to the nearest airport. For the first time in the nation's history, they cleared the skies of all civilian aircraft. Furthermore, they rerouted all the incoming international flights to alternate destinations, many to Canada (Thank You, Canada!) – in many ways, rerouting the flights to alternate destinations was a much more challenging task than simply landing them at the nearest practical airport.

Healthcare Heroes

Healthcare is also an industry of heroes. Since the healthcare industry values its core mission of saving lives, values and activities that strengthen that core mission are celebrated. For example, new drug discovery, new treatment development, service to others, ground breaking research, etc. Celebration of patient safety heroes, however, is relatively new. Some examples of Patient Safety awards include the following:

- John Eisenberg Patient Safety and Quality Award.[26]
- Betsy Lehman Center for Patient Safety and Medical Error Reduction.[27]
- Four Awards from the National Patient Safety Foundation.[28]

The John Eisenberg Patient Safety and Quality Awards are presented in the following categories: (a) Individual Lifetime Achievement for individuals who have demonstrated exceptional leadership and scholarship in patient safety over their careers, (b) System Innovation for projects or initiatives involving successful system changes or interventions that make the environment of care safer, (c) Advocacy for projects or initiatives involving safety-related interventions on behalf of patients, and (d) Research for projects that involve the scholarly exploration of patient safety-related issues.

A review of the profiles of the award winners from 2002–2009 indicates recognition of individual, team, and organizational efforts to build a more positive patient safety culture. Research efforts in understanding medical errors, development and testing of interventions, and system-wide adoption efforts are being recognized by national as well as international awards.

The 2009 award winners[29] are as follows:

- Gary S. Kaplan, M.D. of the Virginia Mason Medical Center, Seattle, Washington, for guiding Virginia Mason Medical Center through a transformation that explicitly placed the interests of the patient first.
- Tejal Gandhi, M.D. of the Brigham and Women's Hospital, Boston, Massachusetts, for increasing knowledge and awareness of safety issues in the outpatient setting and in designing improvement strategies for this setting, particularly through the use of information technology.

---

26 <http://www.jointcommission.org/PatientSafety/EisenbergAward/>.
27 <http://www.mass.gov/dph/betsylehman>.
28 <http://www.npsf.org/npsfac/awards-10.php>.
29 <http://www.jointcommission.org/PatientSafety/EisenbergAward/ 2009+John+M.+Eisenberg+Patient+Safety+Award+Recipients.htm>.

- Michigan Health and Hospital Association (MHA) Keystone Center for Patient Safety and Quality of Lansing, Michigan, for a quality improvement collaborative to focus on interventions to improve patient safety and prevent harm in intensive care units (ICU). More than 1,800 lives have been saved, more than 129,000 excess hospital days avoided, and more than $247 million healthcare dollars saved in the five years since the interventions were first implemented.
- Mercy Hospital Anderson of Cincinnati, Ohio, for developing and implementing a Modified Early Warning System (MEWS), a simple scoring system that is applied to the physiological vital signs routinely measured by nurses.

The *Betsy Lehman Patient Safety Recognition Award* is presented to individuals or organizations who demonstrate leadership and innovation focused on public awareness and education and the development of systemic solutions to medical errors. Each year a particular theme or specific focus in patient safety is selected and tied to the award. The theme of the 2009 award highlighted those who could best demonstrate a culture of safety throughout their organization/facility with specific respect to healthcare-associated infections (HAIs). The award winners were as follows:

- The Division of Cardiac Surgery, UMass Memorial Medical Center for its team's efforts to successfully reduce HAIs through system wide planning, collaboration, communication, and evidence-based actions.
- Marlborough Hospital for encouraging a culture of safety and implementation of several hospital wide initiatives to improve in the reduction of HAIs.[30]

The National Patient Safety Foundation presents four awards: The Socius Award, the Chairman's Award, the Stand Up for Patient Safety Management Award, and the Pfizer Health Literacy in Advancing Patient Safety Award. The Socius, or 'partner' in Latin, Award recognizes the patient safety efforts that stem from a positive partnership between the patients/families and the care providers. The NPSF Chairman's Medal is awarded in recognition of emerging leadership in patient safety. The Management Award is granted to a member hospital of the National Patient Safety Foundation's Stand Up for Patient Safety™ programme in recognition of the successful implementation of an outstanding patient safety initiative that was led by, or created by, mid-level management. The Pfizer Health Literacy in Advancing Patient Safety Award recognizes an individual, group, or

---

30 <http://www.mass.gov/?pageID=eohhs2terminalandL=5andL0 =HomeandL1=ConsumerandL2=Community+Health+and+SafetyandL3=Patient+SafetyandL4=Patient+Safety+Awardsandsid=Eeohhs2andb=terminalcontentandf=dph_patient_safety_c_patient_safety_award_09andcsid=Eeohhs2>.

organization that has made significant strides in improving the health literacy of their constituents or community through research, advocacy, or programme implementation in order to improve patient safety and quality of care.

The 2009 award winners were as follows:

- The Socius Award:[31] MCG Health of Augusta, Georgia received the Socius Award for 'the ground-breaking research work conducted at its Center for Patient- and Family-Centered Care'.
- The NPSF Chairman's Medal:[32] Heidi B. King, acting director of the Defense Department's Patient Safety Program. 'King's efforts have resulted in the internationally-successful TeamSTEPPS™ (Team Strategies and Tools to Enhance Performance and Patient Safety), an evidence-based system aimed at optimizing patient outcomes and promoting a culture of team-driven care. In partnership with the Department of Health and Human Services Agency for Healthcare Research and Quality, TeamSTEPPS was developed to establish interdisciplinary team training systems to serve as the foundation for patient safety strategy. Since its launch in 2006 in the public domain, over 1700 trainers/coaches in the DoD and 930 master trainers through the AHRQ TeamSTEPPS National Implementation Project are assessing, planning, training or sustaining TeamSTEPPS in their organizations. From October 2007 to February 2009, the collaborative effort has grown to astounding numbers – an estimated 35,000 trained and in varying phases of implementation.'
- The Stand Up for Patient Safety™ Award:[33] Mariners Hospital, part of Baptist Health South Florida. Mariners formed a multi-disciplinary team to implement a set of safety measures to identify patients at high risk of developing contrast-induced nephropathy. Their interventions resulted in 'identification of 170 patients with an increased risk of developing this potentially fatal complication, and the program also aided in the identification of nine patients in stage four of chronic kidney disease and three patients in stage five.'
- The Pfizer Health Literacy Award:[34] Dr. Michael Wolf, Director of the Center for Communication in Healthcare at Northwestern University's Feinberg School of Medicine, and Dr. Rima Rudd, Senior Lecturer on Society, Human Development, and Health at the Harvard School of Public Health.

---

31 <http://www.npsf.org/pr/pressrel/2009-5-21_1.php>.
32 <http://www.npsf.org/pr/pressrel/2009-5-21_2.php>.
33 <http://www.npsf.org/pr/pressrel/2009-5-22_1.php>.
34 <http://www.npsf.org/pr/pressrel/2009-5-22_2.php>.

Labour Union Perspective

In aviation, labour unions have played a significant role in improving flight safety. For example, the Air Line Pilots Association (ALPA), which is the largest pilot union in the world, has been actively engaged in advocating, and participating in, safety improvement initiatives since the early 1930s. Some of the key contributions attributed to ALPA's leadership[35] are the following:

- 1933: limited flight time for pilots and copilots to 85 hours per month.
- 1940: started advocating an independent air safety board, which was realized in 1966 with the creation of the National Transportation Safety Board.
- 1944–1947: worked with international partners to lobby for, and eventually create, the International Civil Aviation Organization.
- 1948: partnered with British and Canadian organizations to build the International Federation of Air Line Pilots Association (IFALPA) as the multinational pilots organization; used the IFALPA committee structure to promote flight safety around the world.
- 1953: held the first annual Air Safety Forum.
- 1954: proposed pilot immunity in reporting programmes – the first step toward the creation of Aviation Safety Reporting System (ASRS), Aviation Safety Action Program (ASAP), and Flight Operations Quality Assurance (FOQA) Program.
- 1972: 30 years of lobbying efforts from ALPA contributed to the creation of the Airport Certification programme, which requires airport operators to be certificated by the FAA.
- 1976: as a result of over 20 years of lobbying efforts by ALPA, NASA began operating the ASRS system.
- 1989: several years of active participation from ALPA culminated in the release of the Crew Resource Management Advisory Circular by the FAA.
- 1991: the FAA established Aviation Rulemaking Advisory Committees (ARACs) to bring industry and government representatives together and jointly develop new regulations that will be supported by the industry. ALPA has played a key role in all the ARACs.
- 1995: ALPA helped launch the ASAP and FOQA programmes.
- 1999: FAA announced 'One Level of Safety' to bring commuter airlines as well as cargo airlines to the same safety standards as major passenger airlines. ALPA has been lobbying for this rule since 1995.
- 2004-2010: ALPA is working with airlines, FAA, and the research community to build an appropriately de-identified, but shared, database of ASAP and FOQA data so that systemic issues may be addressed in a proactive manner.

---

35   <http://www.alpa.org/AboutALPA/OurHistory/tabid/2235/Default.aspx>.

Mechanic unions like the International Association of Machinists and Aerospace Workers (IAMAW),[36] Aircraft Mechanics Fraternal Association (AMFA),[37] and Transportation Workers Union (TWU)[38] are also actively engaged in flight safety championship. Some examples of their contributions are as follows:

- Participating in the development of Maintenance Resource Management (MRM) training programmes, ASAP programmes, incident/accident investigations, and general promotion of safe behaviours.
- Promoting professionalism and sharing best practices across airlines through various media – conferences, newsletters, flight safety committees, etc.
- Participation in industry working groups on topics like fatigue, suspected unapproved parts, deicing, aircraft fuel tank safety, etc.
- Participation in research projects concerning maintenance safety – effectiveness of MRM and ASAP programmes, safety ethics, Maintenance LOSA, development of new training/inspection tools, etc.

The National Air Traffic Controllers Association (NATCA) represents over 20,000 members, which include air traffic controllers, certain engineers, staff, and traffic management coordinators. Flight safety is NATCA's prime concern and therefore they participate in a wide variety of technical and operational committees. According to NATCA, air traffic controllers aim to serve the flying public 'with perfection as its minimum acceptable level of performance'.[39] Similarly, the Professional Aviation Safety Specialists (PASS)[40] union represents approximately 11,000 FAA and Department of Defense employees who work on the National Airspace System equipment. PASS's motto is 'we certify air safety'. Clearly, safety of equipment, which in many cases means reliable and accurate operation of the ground-based navigation, communication, and surveillance equipment, is the focus of PASS membership.

As evident from the above discussion, the pilots union are much more strategic in their advancement of safety goals. They have the membership strength, the technical knowledge and the political skill to be involved in lobbying activities as well as research activities that keep them in the lead role. The mechanics' unions, the controllers' union, and the technical specialists' union are more tactical in their pursuit of safety initiatives. Nonetheless, there are many occasions when all the unions, corporate management, and the FAA work together to address priorities of national importance.

---

36 <http://www.iam141.org/members/safety.html>.
37 <http://amfanational.org/>.
38 <http://www.twu.org/index.php/international/our_union>.
39 <http://www.natca.net/about/default.aspx>.
40 <http://www.passnational.org/>.

Two Examples of Nationwide Implementation of Safety Strategies

Example 1: Ascension Health

Ascension Health is the United States largest Catholic healthcare system with 113,000 associates serving across more than 500 locations in 19 states and the District of Columbia. Ascension was created on November 1, 1999. Over 10 years, several different health systems with their own organizational cultures have been added to Ascension, and integrating all these independent cultures into one homogenous Ascension Health has been a major challenge. However, through dialogue and concerted efforts, they focused on a *Call to Action* to challenge all the associates toward three promises that would define Ascension Health:

- To create a sustainable, holistic healing environment that is safe, accessible, appropriate, adaptable and affordable.
- To eliminate all preventable injuries and deaths.
- To create 100 per cent access to healthcare in communities that Ascension Health serves.

In a series of 10 articles[41] in the *Journal of Quality and Patient Safety*, several authors chronicle Ascension Health's journey toward the bold and incredible goal of 'zero preventable injuries and deaths by July 2008'. Pryor, Tolcin, Hendrich, Thomas, and Tersigni (2006) call this journey 'transformational' because it represents more than incremental change; it requires 'changes to the very environment in which care is provided', and it calls for not 'what conventional wisdom holds as the maximum that can be incrementally achieved. Rather, it is led by a vision of what *should* be and a goal to achieve that vision.' In October 2002, at a retreat of 120 Ascension Health leaders, a call to action was developed in the form of a promise to collectively provide, '*Healthcare that works, healthcare that is safe, and healthcare that leaves no one behind.*' All three were 'Big Hairy Audacious Goals' (Collins and Porras, 1997), guaranteed to challenge all the leaders across the Ascension Health system as well as to unify the diverse organizational cultures from all the different independent units that had merged into one Ascension Health with a brand new set of goals that would define Ascension Health forever.

Ascension Health uses a distributive leadership model that provides sufficient autonomy to the leaders of individual hospitals and hospital systems that are distributed throughout the nation. Therefore, in order for the above stated goals to be actionable, they had to be translated into measurable goal statements that made sense to the leaders of the respective hospitals and hospital systems. In December 2002, the leadership team developed a definition for *healthcare that is safe* as 'excellent clinical care with no preventable injuries or deaths'. Next, the team built

---

41 All ten articles are available at <http://www.ascensionhealth.org/index.php?option=com_contentandview=articleandid=26andItemid=139>.

a detailed vision document called Destination Statement II, which builds on a prior visions document, (Destination Statement I). The new document laid out seven aims, 10 rules, five challenges, and three sets of metrics to measure progress. Of these, it is important to note that six of the seven aims were identified by the Institute of Medicine (Crossing the Quality Chasm report, 2001), and one of the ten rules explicitly stated, 'safety as a system property'. The five challenges were: culture, business case, infrastructure investments, standardization, and working together. In this context, Ascension Health decided to focus its cultural challenge on two aspects: teamwork and making safety an institutional priority. By focusing on teamwork, the goal was to shift the mindset among healthcare providers – from 'it is the decisions I make that determine outcomes' to 'I am responsible, at least in part, for creating a safe healthcare environment for my patients and for many team members'. By making patient safety an institutional priority, it was reflected in the goals, reward systems, and measurement. Further, stating 'safety as a system property' in one of the rules strategically shifts the focus of all error investigations to systemic issues. Ascension Health adopted the Safety Attitudes Questionnaire as their survey instrument to regularly assess the safety climate and to tie culture to incentives within the system, elevating patient safety to the overriding priority in implementing innovative solutions. Next, it is interesting to note the three sets of metrics: long-term, short-term, and process. Mortality rate was used as the long-term metric, decreases in preventable incidents such as cardiac arrests as a result of rapid response teams was considered to be an example of a short-term metric, and degree of participation in the safety climate survey was considered to be a process metric. They further refined these metrics, particularly the long-term metric to mean, 'We will consider preventable mortality eradicated in July 2008 if we achieve a 15 per cent reduction in mortality among patients not admitted to the hospitals for end-of-life care'. Achieving this goal would save 800 lives.

The Destination Statement II was first developed in January 2003 and it was approved by their Board of Trustees in July 2003 and handed off to a rapid design team in September 2003. The design team was charged with developing the implementation strategy. To achieve the vision for *healthcare that is safe*, the team recommended that Ascension Health focus on eight priorities for action. These priorities were validated in October and November 2003. Moving forward, it was decided that all the hospitals would focus on three common priorities: Joint Commission National Patient Safety Goals and core measures, preventable deaths, and adverse drug events. Then, certain sites were selected for the remaining priorities as 'alpha' sites or key sites that had the greatest potential for successful implementation, creating a set of best practices and collection of tools ('the change package') that could be adopted across the system. The Institute for Healthcare Improvement's (IHI) *100k Lives Campaign* validated Ascension's priorities – IHI's priorities mapped very well with Ascension's activities.

The results of these efforts were astounding. Compared to the baseline data collected between April and June 2004, the April–June 2005 data indicate 21 per cent reduction in mortality rate, 'exceeding the July 2008 target and corresponding to 1,200 deaths prevented'!

Hendrich, Tersigni, Jeffcoat, Barnett, Brideau, and Pryor (2007) present a reflective review of Ascension's journey and the lessons learned. The key points are as follows:

- The Ascension leadership (all the CEOs and their Boards) were fully committed to patient safety. They were willing to commit resources to achieve the necessary changes in the structures, processes, and performance of their organizations (since Ascension has a distributed leadership model, each facility has its own CEO and Board of Trustees).
- The goal of zero preventable injuries and deaths called for a transformative change, which started with a Call to Action.
- The Call to Action was inspired from the organizational values of Ascension Health and established a sense of urgency – there was a definite goal with a definite timeline; both were publicly announced.
- The creation of a guiding coalition was consistent with the values and culture at Ascension Health; it gave appropriate opportunities for the experts at various facilities to come together and collaboratively develop the Destination Statement II, which was essentially a vision statement with clear priorities for action.
- The priorities for action allowed everyone to focus on the eight elements and align resources as well as evaluation metrics to measure organizational performance in accordance with the eight priorities for action. These priorities also aligned very well with national priorities in patient safety; consequently, there was external validation.
- Internal as well as external benchmarks were used to objectively and transparently evaluate the progress along each of the priorities.
- Incentive pay for the care providers and managers was tied to the achievement of the patient safety goals.
- Several different means of communication were used to train the care providers, to communicate success, and to share best practices across facilities.
- The implementation strategy involved the use of alpha sites, or key implementation sites, for specific priorities. Each alpha site volunteered for the specific priority and was selected after a review. This approach harnessed the pre-existing enthusiasm for specific initiatives and created examples of early success that could be shared across the system.
- There was vertical and horizontal transparency and accountability: all the way up and down each healthcare facility as well as across all the facilities, everyone knew the performance goals, the status of each goal, and the tactics to be implemented to achieve those goals.

Moving forward, Ascension Health has committed to 'Healing without Harm by 2014', a new campaign to continue to improve healthcare quality and patient safety, with a goal of becoming a high-reliability organization in five years. 'As a high-reliability organization, Ascension Health will support its holistic care mission by linking shared values and beliefs with behaviors' (Ascension Health Annual Report, 2009). This commitment is the ultimate expression of our Safety Culture Pyramid!

Example 2: Safety Management System in Aviation

As discussed previously, Safety Management System (SMS) is a means to manage safety performance by managing the underlying processes, much like the quality management programmes. The primary reason for implementation of SMS programmes in the United States is compliance with the ICAO requirements and international harmonization. It is generally believed that since the aviation accident rate is already extremely low, the direct effect of SMS on the overall accident rate will be relatively small, but it will enable the industry to continue to grow without increasing the frequency of accidents. While small operators may not be eager to implement an SMS system, large operators believe that they can leverage existing quality and safety systems/programmes to provide a more unified approach to safety risk management.

Since the Federal Aviation Administration (FAA) serves as the umbrella governing body addressing equipment certification requirements, personnel certification and training, design and manufacturing of equipment, operational requirements, and maintenance requirements, FAA often takes the lead in encouraging the industry to adopt new safety initiatives. The industry, on the other hand, may be somewhat hesitant to voluntarily adopt additional procedures or constraints and add to its operating cost, especially if all the operators in that market segment are not equally impacted. Airlines operate at a very low profit margin, if any. Therefore, it is critical for every airline to reduce its operating costs. Because of this, airlines are generally in favour of system-level mandates or completely voluntary measures. At this time, SMS adoption is completely voluntary.

The SMS Implementation Guide[42] provides the basic structural layout and some procedural guidance regarding the various elements of an Aviation Safety Department that will be entrusted with developing and implementing the SMS. Structurally, this is a top-down approach. The CEO has to commit to supporting the SMS programme and must invest resources in staffing as well as addressing the discrepancies identified by the reporting system. A vice-president level position should be designated for the day-to-day oversight of the SMS programme and for reporting of SMS results to the executive board. This VP then builds his/her team

---

42 SMS Implementation Guide, Revision 2. Retrieved from <http://www.mitrecaasd.org/SMS/documents.html>.

to form the Aviation Safety Department, including liaisons for each business unit (flight operations, maintenance, dispatch, etc.). These structural elements establish safety as an organizational priority, and appoint people across the organization to become champions of safety and to promote a positive safety culture. Further, they establish clear procedures for collection, analysis, and dissemination of safety data/performance outcomes.

At the heart of this system is the Safety Risk Management process[43] that obtains data from multiple sources and in multiple forms, integrates the data, analyzes the data, develops appropriate controls, and monitors the effectiveness of the controls. Depending on the organization's technical capacity to process data, it is possible to for the organization to move from reactive to proactive and from proactive to predictive safety management.

Hazards are identified through a myriad of reporting mechanisms (ASAP, FOQA, Internal Audit, Accident/Incident Investigation, regulatory violation, etc.). All the reports are received by the Aviation Safety Department and depending on the urgency and severity of the hazard a determination is made as to whether or not to invoke the 'fast track' process. Under normal circumstances, the data flows to the Integrated Safety Round Table (I-SRT), which consists of representatives from all the appropriate business units, ensuring a holistic review of the hazards. Assuming that the reported hazard is of interest to the I-SRT, it is assigned to the appropriate division(s) for safety risk assessment. A pre-established matrix is used to determine whether or not the risk is acceptable. The safety risk assessment matrix incorporates safety as well as financial metrics. If the risk is greater than acceptable, an internal data collection effort is launched to validate the hazard and determine the causal factors. According to the 'just culture paradigm', it is recommended that 80 per cent of the intervention be focused on systemic changes. Once the systemic safety risk controls are developed a financial cost-benefit analysis as well as substitution risk assessment (are new risks introduced in the system?) is conducted before the control is implemented. Once the control is implemented, the I-SRT is informed and a follow-up schedule is established.

At any given time, an airline may be reviewing hundreds of hazards. The SRM process described above does not include external validation of the hazard. For example, is this hazard noted by another airline? Another airline may have noted the hazard and implemented a solution. Similarly, solutions implemented by one airline may be shared across multiple airlines without each airline having to go through its own hazard identification and resolution process. Once the SMS system is implemented and has reached sufficient level of maturity, it may be possible for airlines to further enhance the system by linking their systems through the Aviation Safety Information Analysis and Sharing (ASIAS) system.

---

43   Sample SMS Manual, p. 12. Retrieved from <http://www.mitrecaasd.org/SMS/documents.html>.

Chapter Summary

Safety strategies are built to achieve a particular safety goal. Leaders play a pivotal role in establishing the safety goals and driving the specific safety strategies. While most visible examples of safety leadership are from people in formal leadership roles, there are noteworthy examples, particularly in healthcare, where the leadership has come from patients and their families, transforming the way healthcare handles errors. Alignment across organizational mission, values, strategies, structures, processes, and practices is critical in achieving a strong safety culture. In the aviation industry, the Federal Aviation Administration is typically involved in providing leadership and unifying guidance materials for safety strategies that may be implemented across different types of aviation organizations. In the healthcare industry, in the absence of a unifying regulatory agency, multiple national organizations have taken leadership roles and developed appropriate standards and training materials. In both industries, heroes and legends have played an important role in shaping the culture. Today, a review of the accomplishments of these individuals and the criteria for awards and recognition by the two industries provides a valuable insight into the behavioural characteristics as well as performance metrics that are espoused and admired by the two industries.

Finally, the examples presented in this chapter provide a quick overview of how the two industries are developing safety strategies to improve the safety culture in their respective domains. Both examples illustrate national-level strategies, but one from a corporate perspective and the other from a regulator perspective. Although one might find aviation leading the healthcare industry in many aspects of safety, we still don't have one airline articulating and implementing a safety goal like Ascension Health. On the other hand, while Ascension has done an excellent job of improving its safety culture, few other hospital systems have declared a goal of 'zero preventable injuries and deaths'.

References

Ascension Health Annual Report (2009). Integrating excellence and stewardship: 2009 ascension health annual report. St. Louis: Ascension Health. Available from <http://www.ascensionhealth.org>.

Chirinos, F. (2008). Spohn probes medicine error: 14 babies were given too much blood thinner. Retrieved from <http://www.caller.com/news/2008/jul/08/spohn-probes-medicine-error/> on April 25, 2010.

Collins, J.C. and Porras, J.I. (1997). *Built to Last: Successful Habits of Visionary Companies*. New York: HarperCollins.

FAA (2006). AC 120-92 Introduction to safety management systems for air operators, AFS-800. Available at <http://www.faa.gov>.

Hendrich, A., Tersigni, Al., Jeffcoat, S., Barnett, C., Brideau, L. and Pryor, D. (2007). The ascension health journey to zero: lessons learned and leadership perspectives. *The Joint Commission Journal on Quality and Patient Safety*, 33(12), 739–749.

INPO (2004). Institute of nuclear power operators: principles for a strong nuclear safety culture. Retrieved from <www.efcog.org/> on April 25, 2010.

McVenes, T. (2007). Safer skies: we are almost there. *Air Line Pilot*, 14–19. Air Line Pilots Association.

NTSB (1992). United Airlines flight 232 McDonnell Douglas DC-10-10 Sioux Gateway Airport, Sioux City, Iowa, July 19, 1989. Aviation accident report number AAR-90/06. Retrieved from <http://www.ntsb.gov/publictn/1990/AAR9006.htm>.

Patankar, M.S. (2009). A practical guide to maintenance ASAP programs. Report prepared for the Federal Aviation Administration (FAA Document DOT/ FAA /AR-09/28). Available from <http://hf.faa.gov>.

Patankar, M.S. and Taylor, J.C. (1999). Corporate aviation on the leading edge: systemic implementation of macro-human factors in aviation maintenance. In: *Proceedings of the SAE Airframe/Engine Maintenance and Repair Conference* [SAE Technical Paper Number 1999-01-1596]. Vancouver, BC.

Pryor, D., Tolcin, S., Hendrich, A., Thomas, C., and Tersigni, A. (June, 2006). The clinical transformation of ascension health: Eliminating all preventable injuries and deaths. *The Joint Commission Journal on Quality and Patient Safety*, 32(6), 299–308.

Steitzer, S. (2009). Jury awards $7.1M in Comair crash lawsuit. Retrieved from <http://courier-journal.com> on April 25, 2010.

Taylor, J.C. and Patankar, M.S. (2001). Four generations of MRM: evolution of human error management programs in the United States. *Journal of Air Transportation World Wide*, 6(2), 3–32.

Chapter 6
Safety Values

Introduction

Shared values and beliefs are the foundation of a culture. In understanding the safety culture of an organization, it is critical to delve deep into the discovery of the shared values, beliefs, and unquestioned assumptions. Geert Hofstede's (1984) scales of individualism versus collectivism and power distance have been used to describe differences in cultures based on organizational, national, and professional groups (Helmreich and Merritt, 1998; Patankar and Taylor, 2004). For example, European cultures are more individualistic than those of Asia, which tend to be more collectivistic. There's also a higher power distance between supervisors and subordinates in Asian countries than in European countries. From the perspective of professional cultures, surgeons tend to be more individualistic than engineers (Helmreich and Merritt, 1998); pilots tend to be more individualistic than surgeons (Helmreich and Merritt, 1998); and mechanics tend to be more individualistic than pilots (Taylor and Patankar, 2004). In effect, differences between organizational, professional, and national cultures have been expressed almost entirely in terms of differences in values across these three groups. Our Safety Culture Pyramid model asserts that safety values and unquestioned assumptions influence safety strategies, organizational structures and policies, as well as recognition of heroes and legends, which collectively influence safety climate, and ultimately safety performance. Therefore, in order to fully characterize the safety culture at a given organization, and at a given time period in its evolution, we must discover the enacted values at the base of the Pyramid and define the gaps between enacted values (values that are actually practised) versus espoused values (values that are advertised as a corporate values) (Argyris and Schön, 1974, 1978). Next, in order to transform a culture, or under circumstances of organizational mergers and acquisitions, special attention needs to be focused on acculturation efforts so that the desired cultural transformation is achieved.

In this chapter, we present two techniques that may be used to discover enacted values: deep dialogue and narrative analysis. These enacted values can then be compared with the espoused values, which are more readily available through corporate marketing materials as well as profession-specific credo. The effects of gaps between espoused versus enacted values as well as strategies to align the two are discussed. Since we use Rochon's (1998) definition of transformation, for a cultural *change* to be considered cultural *transformation*, it has to be deep enough that it is born from value conversion, value connection, or value creation.

We conclude the chapter with a discussion of acculturation (cf. Chun, Organista and, Marín, 2003) and value-based transformation that is motivated by patients.

Value Discovery: Deep Dialogue and Narrative Analysis

As discussed in Chapter 2, dialogue is an intense conversational technique that can be used to uncover underlying values and unquestioned assumptions that are held by individuals, among professions (e.g. surgeons, pilots, engineers, etc.), as well as organizational units (e.g. flight operations versus maintenance or emergency room versus pharmacy). The primary purpose of a dialogic conversation is to engage people in productive and reflective thinking so as to fully express the deepest beliefs and unquestioned assumptions that may be linked to their behaviours. Since most professionals in high-consequence industries are not trained in the art and science of dialogic communication, and certain teams may have particularly caustic relationships, it may be easier to bring in an external facilitator. Such a facilitator, however, needs to have professional credibility (from a similar professional background or particularly known for safety enhancements through dialogic communication in other organizations/industries). If a senior member of the existing team wants to take on this challenging task, the person needs to be open to criticism by the team members as well as committed to changing his or her own behaviours – the team leader needs to model reflective practice. A meaningful dialogue can occur only in a safe environment where there is no fear of reprisal, there is interpersonal trust between the people expressing their feelings and experiences and the person listening and in charge of implementing changes, there is acceptance of responsibility for past errors and commitment to prevent them in the future, and there is a commitment to be non-judgmental and non-punitive. Such an environment fosters interpersonal trust and facilitates a meaningful dialogue.

Abma and Widdershoven (2005) connect narrative, storytelling, and dialogue with evaluation methodologies to discover the value and meaning of professional care (e.g. nursing) and as a way to learn, understand and improve clinical practice. They present examples of personally sensitive stories from nurses about stressful situations. Issues identified from such narratives can be used in dialogic situations to stimulate further reflection and learning. They describe their dialogical approach as characterized by 'listening, probing, and deliberation' rather than 'confronting, attacking, and defending' (p. 95). Dialogue allows for the surfacing of assumptions and the mutual construction of meaning for significant but ambiguous events. To prevent bias and establish validity, this method frequently uses triangulation of sources and inclusion of different perspectives.

Chris Argyris (e.g. Argyris and Schön, 1974; Argyris, 1995) describes a useful technique that can be applied to dialogue to uncover disconnects between what is publicly stated and what is privately thought and felt during communication about important topics. Argyris refers to this technique as 'the left and right hand column case method'. This method involves asking a person to write a brief case that begins with a statement of a problem, includes a problem-solving strategy, presents the conversation that the case-writer would have with another person, and also states what information and feelings the case-writer would not share with the other person. The narrative of the case is divided into a right column (what is actually said to another person) and a left column (private thoughts and feelings that remain unshared in the conversational interchange). This method can reveal unstated motives, untested assumptions, hidden belief systems, knowledge sharing reluctance, influence tactics, and other influences on public behaviour. The 'left hand column' of the case affords an opportunity for reflection, insight, learning, improved dialogue, interaction, and performance.

Narrative analysis, on the other hand, is a process of eliciting stories or experiences of employees across the organization to extract themes that can be associated with enacted values. Meyer (2003) presents a case example from a child care organization at which he asked two research questions:

- What values are embedded in stories told by an organization's members?
- Are the values consistent or inconsistent with one another?

Meyer used structured interviews to collect the stories or narratives, which were defined as 'any sequence of events together in time or causally related, with organization-related characters, which takes place in a setting somehow related to the organization'. He collected 555 stories and categorized them by major value, and prioritized these values, as indicated by the participants:

- Consideration
- Organization/planning
- Timely information
- Participation in decision-making
- Discussion of conflict
- Friendliness
- Clarity of messages
- Commitment
- Autonomy
- Authority.

In a story collection, both positive as well as negative stories may be associated with the same value. For example, while some people might tell stories that revolve around examples of care and consideration, others might tell stories about *lack* of consideration. Both groups, however, value *consideration*. Similarly, Meyer notes that values such as autonomy and authority may conflict with each other at times, particularly when they are ranked similarly (No. 9 and No.10, respectively). Meyer discovered that while 'those in authority were valued and respected, the rules imposed by authorities were seen as needlessly interfering with staff autonomy and ability to make choices'. In high-consequence industries, the most commonly contested values are safety and productivity; wherein, a typical comment is, 'we value safety as long as it doesn't conflict with production'. Meyer expressed an important limitation of his study: the narratives obtained were not spontaneously told to coworkers or the researchers; they were in response to specific questions. Hence, the narratives may represent only a limited sample.

Case Example 6.1: Arlene, a Social Worker at a Large Community Hospital

This case was initially presented by Patankar, Brown, and Treadwell (2005, Chapter 6). It is used here to illustrate how retrospective analysis can be done to identify disconnects in values – both across professions and between espoused values versus enacted values.

Arlene is a social worker in a large community hospital. She has served as a social worker for 29 years. She reports the following experience:

> My work includes helping hospitalized patients prepare for post-discharge care. I have a story that is typical of the kind of dilemma I deal with every day. An elderly patient had been admitted for a surgical procedure. Her plan was to return to her apartment in an independent living facility upon discharge from the hospital. After her surgery those of us who were working directly with her noticed that she was not the same lady – not quite right cognitively, and physically very weak. Because of her weakened condition I asked the surgeon if she could have some skilled aftercare before discharge to her home. He said no, that her affect and demeanor were 'consistent with her baseline (i.e. her pre-procedure assessment)'. The surgeon visited with the patient for about five minutes on rounds in the morning, matched the patient's appearance and condition to his expectations, and then departed. He didn't take time to really assess her condition, or test his assumptions. He says he puts patient care first – his website says something like 'the quality of your care is my foremost concern', but his words are not consistent with his actions.
>
> Meanwhile, both the physical and occupational therapists had seen the patient and said that she could not function on her own – that it took both of them to assist her in and out of bed. We paged the surgeon again. Over the phone, after hearing what the physical therapist had to say, he said that she could be discharged to the rehabilitation facility, but none of us felt that she was ready for this. We recommended that she be discharged to a skilled nursing facility. He said no, that she would be fine in rehabilitation. I was very worried at this point and asked the surgical service's nurse practitioner to examine the patient. She said there was no doubt that the patient was not well enough to be discharged to the rehabilitation facility. She paged the surgeon and told him that if he wanted to discharge the patient to the rehabilitation facility he'd have to do it himself. He then came to see the patient, finally agreeing to have her discharged to a skilled nursing facility.

Analysis: The Theory of Action Perspective

The theory of action approach (Argyris and Schön, 1974, 1978) starts with the basic assertion that there is practice in theory and theory in practice. In a sense, this is simply restating the axiom 'do as I say, not as I do'; in other words, there are gaps or discrepancies between what each of us does, as others see it, and how we see it ourselves. These gaps can be sources of interpersonal problems, lack of coordination in organizations, or simply a matter of humour that is of relatively little consequence in day-to-day affairs. But Argyris and Schön go a step farther. They assert that there is an implied logic of action that drives our behaviour; it is largely tacit, taken for granted, and invisible to us because we act *from* it rather than being consciously aware of it most of the time. Only when our logic of action gets us in trouble do we question it. This usually happens when we are surprised, frustrated, or disappointed in the outcome of our actions. But these are also moments that typically heighten our defences, since the unintended consequences of our actions usually attract more attention than those that are intended and we are 'caught' making a mistake, disappointing others, or falling short of our expectations for ourselves. These are moments which call for learning but produce anti-learning responses, such as avoidance, cover-ups, deflection of the conversation, and point-counterpoint arguments. The work of Arygris and Schön explores ways in which taken-for-granted assumptions and unrecognized values can be surfaced and examined in ways that lead to learning and can create powerful leverage for desired changes in culture, practices, and operating assumptions.

Let us look at the theory of action of the surgeon, by examining his actions, or behaviours as described by Arlene. These include the following:

- The surgeon relied heavily on his own assessment of the patient, initially discounting Arlene's report and recommendation, as well as recommendations of physical therapists and occupational therapists.
- The surgeon eventually agreed to discharge the patient to a rehabilitation facility, but not to a skilled nursing facility, as requested by the social worker and therapists.
- The surgeon agreed to reassess the patient only after the surgical service's nurse practitioner refused to discharge the patient to either her home or a rehabilitation facility.
- Finally, the surgeon agreed to discharge the patient to a skilled nursing facility after being compelled to re-assess her.

It took a great deal of time and effort on the part of four individuals, in four different disciplines, to get the surgeon to reassess the patient. To start with, let us infer what may be the intended and unintended consequences of the surgeon's behaviour. The unintended negative consequences are many. From Arlene's perspective he has missed diagnostic cues that related to the patient's physical incapacitation and cognitive decline. In doing so, he compromised the quality of care, and reduced patient, family, and staff satisfaction through the poor quality of interaction and decision-making. The surgeon appeared to have made his decisions unilaterally, without consultation with other disciplines involved in the patient's care. Worst of all, from Arlene's perspective, the surgeon missed key signs of the patient's medical status that could put the patient at risk or, at best, place her in a situation which will lead to decline in her condition rather than improvement. Notwithstanding these negative effects, what may be the intended consequences of the surgeon's actions?

The surgeon appeared to have been operating from a perspective that would allow for a high volume of work and significant professional autonomy. In fact, operating as a highly autonomous decision maker enhances his efficiency. He matched the patient's post-surgical state with the pre-surgical baseline, missing indications of deterioration in the patient's condition that had been observed by members of other disciplines, such as the decline in her ability to think, communicate, and move independently. His approach to conducting rounds on his patients may work well on most occasions, but in this instance he had not identified important changes in the patient's health status. From the perspective of team-based practice, the surgeon's approach is flawed: the surgeon seems to operate from a craft model perspective, as a solo practitioner. He is the subject matter expert at the 'top of the food chain' when it comes to making decisions about the patient's care. From his perspective, this seems to be sufficient; from Arlene's perspective, it is woefully lacking. The conditions of clinical practice that create the conditions described by Arlene are not new. A theory of practice that emphasizes the decision-making autonomy of physicians has long existed and is instilled even today in medical education. The organizational structure of hospitals was established to support this, as we will discuss in the next section.

Analysis: The Professional and Organizational Divide

Hospitals are unusual among socially and technologically complex organizations in that they maintain a pre-industrial organizational structure (Sharpe and Faden, 1998; Merry and Brown, 2001). Specifically, they were conceived to support the practice of independent physicians, at a time, early in the twentieth century, when the therapies and care technologies available to physicians were very limited and the practice of medicine was largely a solo endeavour. Few patient care roles were required in hospitals – usually a physician would be assisted by nursing staff and perhaps orderlies, the historic equivalent to today's 'patient care technician'. With rapid advances in medical science and technology over the past 100 years, numerous new technical specializations and patient care roles have emerged. Yet, despite this rise in social and technological complexity, hospitals remain organized fundamentally as they were in 1910, functioning as facilities within which the practice of affiliated physicians is supported by material and human resources. The relationship remains transactional in that physicians gain access, when granted hospital privileges, to the patient care resources of the hospital and, in turn, the hospital receives revenue from patients admitted by physicians through the provision of beds, diagnostic technologies, supplies, nursing care, and other services. However, the division of authority and responsibilities between the medical staff and administrators effectively creates two loosely aligned organizations under the umbrella of a shared Board of Trustees (Figure 6.1).

Hospital administrators are responsible for the financial management of the hospital and for the provision of patient care facilities and resources, but they do not exercise authority over medical practice or medical affairs. Medical staff members (physicians) are accountable for the quality of medical care. Oversight functions such as medical peer review and credentialing are typically overseen by Medical Staff Executive Committees. In the century that has elapsed since this organizational structure emerged, the structural and conceptual divide between medical and hospital functions has remained largely untouched. The autonomy of the medical practitioner is still strongly protected within this organizational model. Medical education has perpetuated a theory of practice in which the physician is taught that she or he is ultimately accountable and responsible for the care of patients and hence is the ultimate decision-maker. While this sense of accountability and responsibility is important and laudable, when coupled with the archaic organizational structure of hospitals, it has perpetuated a climate in which physician autonomy may eclipse the mindful engagement of other care providers, patients, and patients' families. It is a difficult climate in which to align values and goals across disciplines and functional areas.

In Arlene's case, the theory of action perspective suggests that there is logic to the entire situation described by Arlene which is organizational in nature and also that there is logic to the action of the individuals. In this case, the surgeon's unspoken assumptions and underlying values appear to be driven by the invisible,

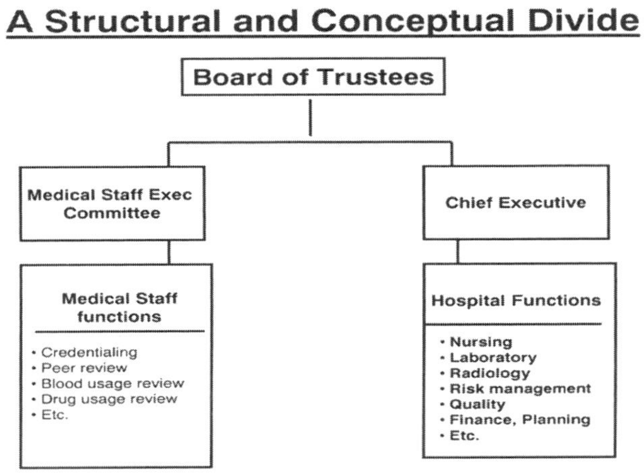

A Structural and Conceptual Divide

Board of Trustees

- **Medical Staff Exec Committee**
  - **Medical Staff functions**
    - Credentialing
    - Peer review
    - Blood usage review
    - Drug usage review
    - Etc.
- **Chief Executive**
  - **Hospital Functions**
    - Nursing
    - Laboratory
    - Radiology
    - Risk management
    - Quality
    - Finance, Planning
    - Etc.

Figure 6.1   **The bifurcate structure of hospitals (Merry and Brown, 2001)**

but powerful, view of role-based hierarchy. One can infer that the views of physical and occupational therapists, and of social workers, do not count as much for the surgeon as the view of a nurse practitioner, whose professional credentials provide a higher level of authority than the social worker or therapists. In addition, it seems that the nurse practitioner would ordinarily sign off on the discharge plan; this gives her the power to compel the surgeon to revisit his decision by refusing to sign off on what she perceived to be an inappropriate discharge plan. Without a level of professional authority comparable to that of the nurse practitioner, Arlene can only influence through data and persuasion. Arlene and her colleagues assert that the surgeon's concept of role hierarchy and decision-making authority (theory in practice) leads him to resist, and likely discount, the decision-making input of those he views as subordinates. Arguably, his difficulty in receiving input that contravenes his initial assessment may reflect his personality as much as his medical training and education. To dwell solely on his unwillingness to recant, however, would distract us from another important consideration – how people make decisions under conditions of time pressure and uncertainty. Returning to the surgeon's incomplete (at least as seen in hindsight) assessment of the patient, there is another lens that should be applied to the case provided by Arlene. With it we will examine another perspective on why values in practice may seem quite apart from the values we espouse.

Analysis: Decision-making and Efficiency

When reviewing the performance of individuals in the context of a specific event, it is common to make judgments about their values, conscientiousness, competence, or intentions. Yet, without an understanding of the time pressure, performance constraints, and goal conflicts associated with task performance in the context of the surgeon's work, judgments regarding values, motives, and competence our assessment will be incomplete and flawed. In hindsight, we could assert that the surgeon valued efficiency on rounds over thoroughness, and attribute the inadequacy of his assessment to arrogance, carelessness, or neglect. However, the trade-off of efficiency and thoroughness is a necessity in time-pressured environments. How people make decisions under such conditions becomes a very important consideration when trying to understand, in hindsight, why an assessment proved inadequate, or an action was inappropriate. The day-to-day reality in dynamic, time-pressured environments is that people are compelled to balance thoroughness in the performance of their work with the concurrent demand for efficiency. The ever present risk is that our balancing act may be, at least from a situational perspective, incorrect; that the things we usually do to succeed will suddenly prove wrong. We are especially vulnerable to surprise if our theory in practice leads us to act autonomously in highly interdependent settings. Healthcare systems may be thought of as *complex cognitive systems* (Crandall, Klein, and Hoffman, 2006), where safety and outcome reliability hinge not only on the knowledge, cognition, and reasoning of individuals, but upon the knowledge, cognition, and reasoning of the many individuals whose patient care activities, perceptions, and actions are highly interdependent. In complex cognitive systems, it is extremely important to establish the expectation and practice of accepting and inviting alternative perspectives and recommendations – to be willing to revise our *theory of situation* if new information or an alternative interpretation indicates (Bolman, 1980). This willingness to recant was not evident in the surgeon's response to members of other disciplines, as described by Arlene. The challenge of uncoupling perception and action relates not only to what may be observed, but to the nature of decision-making and the human orientation toward efficiency.

The Efficiency Thoroughness Trade Off and Decision-making Heuristics

Hollnagel (2009) describes the balancing act between efficiency and thoroughness as the Efficiency-Thoroughness Trade-off (ETTO) principle. This principle may be viewed as an extension of the Speed-Accuracy Trade-Off (SATO) discussed in Chapter 4. Speed of activity and throughput is an observable manifestation of time pressure and efficiency goals. Underlying the ETTO principle, like SATO, is the assumption that when individuals engage frequently in any given task they will develop the most efficient approach to performing it, as long as they perceive that desired results can be achieved without compromising the quality or safety of the activity and outcome. At the cognitive level, efficiency manifests in the application of unconscious decision-making shortcuts, which may not be easily inferred from the individual's observed behaviour. These are highly automatic features of decision-making that are continuously exercised by all of us as we navigate our daily activities. Decision-making shortcuts may only be made explicit when they result in a mistake – an unintended outcome. They are triggered by recognition of cues and patterns that prompt prior responses from our 'mental library' of experience. Consider the following example: 'It was a little embarrassing. I drive an hour to work every day along the state road, until I turn off onto Route 3, to get to my office. Yesterday, I should have stayed on the state road because I was taking my wife to the airport, but I wasn't really paying attention and automatically turned off onto Route 3'.

Literature on the psychology of perception suggests that people develop libraries of situations or experiences and use them as references. Referred to as schemata (Brewer and Treyens, 1981), these 'references' provide an efficient way to understand a given situation because we don't need to exert great effort in identifying and reasoning about what the situation connotes and how we should respond. In the example above, the most frequent response to the recognition of the exit to Route 3 – to turn off the state road onto Route 3 – was highly automatic, but inappropriate at certain times.

Another way of describing such phenomena is referred to as *Recognition Primed Decision-making* (Klein, 1989). Theoreticians in the field of *Naturalistic Decision-making* assert that an innate feature of human decision-making is that the decision maker will match the features of the current situation to a similar situation (prior experience), and then conduct a mental rehearsal of a possible course of action (Klein, 1993). If the action seems viable, then the decision maker will execute that course of action. Depending on whether the action affects the situation as intended, the decision maker will revise the theory of the situation and adapt their actions accordingly (see Figure 6.2).

Yet another view of decision-making and efficiency is offered by Tversky and Kaheman (1973, 1974), who proposed that people typically use heuristics, or decision-making shortcuts. Heuristics that may be relevant to the surgeon's decision-making, as described in the case presented by Arlene, include *representativeness*, along with *adjustment* and *anchoring*.

The representativeness heuristic comes to bear when there is commonality between circumstances or when there is similar appearance assumed between objects, people, or situations. The highly experienced surgeon described in Arlene's case had performed the surgical procedure undergone by the patient, several hundred times. The associated patterns of disease, surgical intervention, and recovery were well known to surgeon. It is not inconceivable that he matched the surface features of the patient's condition with that of prior patients, who were healthier at that phase of post-surgical recovery. Finally, adjustment and anchoring describes those situations in which the decision-maker makes an assessment that is biased by an initial impression. Once the surgeon had formed his assessment, he may have become 'anchored' in that initial evaluation.

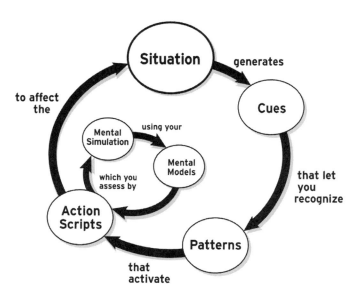

Figure 6.2 The recognition primed decision-making model (Klein, 1989). Recognition primed decision-making model graphic provided courtesy of Gary Klein, Ph.D.

Our objective in visiting the topic of decision-making is to make two points regarding assessment of values in action:

- In the course of real work, under conditions of time pressure, uncertainty, and competing responsibilities, people must continually balance thoroughness in performance of their work with efficiency. Decision-making shortcuts, regardless of how we characterize them, are part of our makeup as human beings. Most of the time our decision-making shortcuts work, or they are quickly and appropriately adapted as our understanding of the situation changes. In socially and technologically complex work settings where there is high consequence for failure, it is imperative that we be alert to signs that our assessments and actions are not working out. And part of being prepared to adapt activity in any given situation is a willingness to receive and consider decision-making input from others. In Arlene's case, the unwillingness of the surgeon to revisit his care plan led other caregivers to perceive that he did not value the provision of high quality care, a perception that is antithetical to his espoused values as a surgeon.
- When people who are working together have conflict regarding perception of the 'right thing to do', it is entirely likely that their values will be invoked, with the potential for heightened conflict. Seeking alignment of values, beliefs, and assumptions is an essential feature of efforts to transform unit-level culture and to overcome conflict that arises at the level of individual and professional values.

In contrast to the situation described by Arlene, which did not yield a change in practice, the following initiative at a major medical centre sought to alter the way clinical personnel interacted with one another by aligning values and goals across disciplines.

Case Example 6.2: Interoperative Pathway for DIEP Flap Surgery

The Deep Inferior Epigastric Perforator (DIEP) flap surgery is breast reconstruction surgery following mastectomy. It involves removal of soft tissue from the abdominal area and transplanting it to reconstruct the breast. It is a very long and arduous procedure. The following example, narrated by a surgeon at a large academic medical centre, illustrates how disconnects in professional values as well as espoused versus enacted values can be addressed through a process called 'intraoperative pathway' which is essentially a communication and process improvement protocol that ultimately enhances teamwork and improves efficiency as well as reduces errors.

After a substantial period of providing the DIEP flap procedures, the team was achieving outcomes that were consistent with published literature. Yet, operating room staff were frustrated with perceived coordination and supply inefficiencies during the DIEP flap procedure. This frustration had escalated to the point where the team held a problem-solving debriefing. They assembled surgeons, nurses, technicians, and Operating Room (OR) leadership to identify the sources of frustration and tried to resolve them. An external facilitator and the perioperative nurse specialist co-facilitated this meeting.

During the meeting, nurses said that there was a lack of clarity with regard to the timing of what they should be doing during the procedure, a concern that was especially strong among those nurses who came on shift when the procedure was well underway. The more experienced nurses often went off-shift before the completion of the procedure and, after eight to 12 hours, verbal communication declines in the OR. It was difficult for incoming nurses to determine the status of the procedure and anticipate their most immediate tasks. The challenge of getting up to speed with the current state of the procedure was amplified in that nurses coming into the procedure for second shift often had less experience with the DIEP flap procedure, which made effective communication of expectations all the more critical. Surgeons were demanding that experienced nurses be available throughout the procedure who could better anticipate what had to happen during the procedure, which late in the day, wasn't always possible. Conversely, nurses were demanding that the surgeons be more communicative, so they could better anticipate needs.

The external facilitator remembers sitting in the room and, as they all talked, he realized that everyone had valid concerns. The real issue was that they didn't have a way for the team to function with multiple personnel substitutions during the procedure. They were trying to solve micro-problems of coordination and support within chunks of the procedure, but the real issue was that there was no structure within which the team could predictably function through the entire procedure. The lack of effective communication and a sense of what nurses and technicians should be doing, and when, was the dominant concern expressed across disciplines. The discussion led everyone to understand that this was the root of conflict within the team; as participants verbalized their frustrations they recognized that the problems experienced by each role were interrelated. The facilitator asked if creating a roadmap would be helpful. Everyone in the room agreed to try it out.

A design group was formed and, as they began to work on the roadmap, it became apparent that another key issue was that supplies were not always available when needed. The procedure would be slowed while awaiting supplies. Although nurses had a supply box for DIEP flap cases, it was often raided during other procedures. So, the team had to determine how to control supplies. In addition, they needed to identify essential competencies for each role by phase of the procedure and to consider how to structure the procedure to support effective team functioning over the entire procedure.

As the team discussed this challenge, the group realized that the problems with intra-procedural delay and problems with coordination might be resolved by mapping the procedure as an intraoperative pathway and identifying key phases of the entire procedure along with tasks, resource requirements, and coordination needs. As a way of thinking about coordination requirements, they adopted Gittell's model of 'relational coordination' (2000) (per Gittell, relational coordination is defined as 'frequent, timely, and accurate communication characterized by problem solving, shared goals, shared knowledge, and mutual respect). This established a conceptual framework for what they wanted to accomplish with the pathway.

The DIEP flap pathway design group divided the flap procedure into nine phases. Tasks and equipment were identified for each role in each phase of the procedure. A locked supply box for the DIEP flap procedure was installed, to ensure that supplies would not be utilized in other cases – a problem which, in the past, had sometimes left the team scrambling for essential equipment and causing delays. Three surgeons worked on the pathway, each agreeing to use the same supplies and equipment, and to aligning their practice with a single pathway to eliminate the variability associated with having had three individual approaches to the procedure.

An initial team briefing was established, and transition from phase to phase included a 'phase briefing', which was guided by a checklist. Supplies were reconciled at the conclusion of each phase to ensure a running count was kept and that any discrepancies were resolved immediately following the surgical phase. In the past, if they waited until the end of the procedure to reconcile, they might have different people in the room (because of nurses and technicians going off-shift), and it could be very challenging to resolve any discrepancies. The team developed the 'Count Safety Checklist' to resolve supply count discrepancies at inter-phase points throughout the procedures.

Another benefit of the inter-phase briefing was that it aligned the entire team on the status of the surgery and any issues or needs that may have come up. It also identified parts of the procedures where surgeons needed to minimize interruption. To prevent change in nursing or tech staff at those points in the procedure, the team created 'no hand-off zones'.

An example would be microsurgery, when the surgeon is focused on the operative field using a microscope. Because this procedure is performed by more than one surgeon, a decision was also made to bring in the second surgeon between phases so that the surgery wouldn't be interrupted. Another thing that surgeons reported is that the primary surgeon will take a break of 10 to 15 minutes prior to beginning the microsurgery phase, and the period of continuous operating time was limited to six hours. To address concerns that nursing and tech staff were sometimes assigned to the DIEP flap procedure without appropriate training, the team identified competencies for each role and phase of the procedure. A core team of nurses and surgical technicians trained to these competencies was established to support the DIEP flap procedure.

The pathway allowed the team to closely monitor the time of each phase of the procedure and to determine how they might improve efficiency. They set benchmarks and when the length of any phase of the procedure deviated from expected, the team would determine if the time efficiency could be improved. All variances were discussed in post-operative debriefings, in support of problem solving and continual improvement. An important point is that the post-operative debriefings have benefitted the team, not only in terms of process improvement, but also in terms of continually aligning around goals and values. The team developed very strong common ground around values, goals, and coordination needs. Any disconnects in this vein were addressed and resolved in team meetings outside the operating room.

Outcomes following implementation of the DIEP Flap Interoperative Pathway were reported in the *Journal of Surgery* (Lee et al. 2008). Some of the highlights are discussed below:

> In the unilateral breast reconstruction group there was a statistically significant decrease in mean operative time of 1.3 hours – a 15.9 per cent decrease from the pre-pathway mean value (8.2 hour pre-pathway to 6.9 hour post-pathway). And the bilateral group had a statistically significant decrease in mean operative time of 2.2 hours. There were cost benefits as well. The mean OR cost per patient decreased by 10.2 per cent and overall hospital cost decreased by 3.6 per cent for unilateral reconstruction procedures. Bilateral reconstruction mean OR cost decreased by 2.0 per cent and overall hospital cost decreased by 1.6 per cent. Complications were analyzed for the immediate postoperative period and there were no statistical differences found between pre and post-pathway groups for return to the OR, total flap loss, partial flap loss, fat necrosis, infection, deep vein thrombosis, and pulmonary embolus. Some additional improvements in quality and safety were achieved however. Before implementation of the pathway, administration of prophylactic antibiotics within 60 minutes of incision occurred 96 per cent of the time. Post pathway, this rose to 100 per cent. The number of patients who received the correct number of repeat dosing of antibiotics, rose from 80 per cent pre-pathway to 92 per cent post-pathway. Timing of repeat doses to occur within 30 minutes of expected dose rose from 29 per cent to 39 per cent. Prophylactic heparin was administered 25 per cent of the time prior to implementation of the pathway. Post-pathway it rose to 83 per cent.

With regard to staff satisfaction following implementation, 82 per cent of respondents perceived that the predictability and consistency of procedures, equipment, competencies, communications, and supplies provided by the pathway enhanced their ability to provide safe and efficient care. 73 per cent believed that the transition guidelines were helpful in their work and 64 per cent of staff perceived that the pathway improved interdisciplinary communication.

An essential pre-condition for the interdisciplinary development of an intraoperative pathway was the facilitated meeting which led to the shared realization that everyone's frustration had common causal factors – coordination problems, absent supplies, and other sources of disruption during the procedure – which were symptomatic of the absence of systematic thinking about the procedure, rather than individual neglect. Ultimately, the frustration expressed by team members stemmed from the absence of a shared mental model of the overall procedure coupled with the absence of explicit situation updates as the procedure progressed. Non-surgical team members were often uncertain about when they should offer situational assistance to the surgeons, or time the performance of expected tasks, either alone or in collaboration with others.

It may be surprising for those not familiar with healthcare that this type of planning and preparation for a new procedure is not the norm. Although interdisciplinary planning and problem-solving meetings are routine in most industries, interdisciplinary meetings among clinical personnel are surprisingly uncommon. Given the absence of opportunity to look at their work from a systems vantage, and to communicate concerns, discuss opportunities for improvement, and learn more about each other's roles and responsibilities, it is not surprising that personnel ended up being frustrated with one another. Interdisciplinary planning and problem-solving provide a powerful means of gaining a shared understanding of the big picture of work for which team members are jointly accountable. One reason for the rarity of interdisciplinary team meetings, especially those designed to harness the knowledge and information resources of all participants, is that education and training for health professionals is oriented toward supporting the autonomous decision-making and action of physicians rather than interdisciplinary practice. This reflects both professional and organizational culture biased toward independence rather than interdependence. The surgeon's sense of what the DIEP Flap team accomplished by developing the intraoperative pathway reflects valuable insight into essential practices for cultural change:

> We have developed an environment in which team processes may thrive, but to do so required dialogue to achieve trust and a sense of shared values and purpose among the team members. The team, through recognition of a shared problem, ceased blaming one another for inefficiencies during the DIEP flap procedure, and began engaging in the development of shared beliefs, values, and assumptions among team members that enhanced predictability of tasks and activity. The DIEP flap pathway, in very tangible ways, may be thought of as an explicit compact to provide mutual support.

An examination of the DIEP Flap team's development of the intraoperative pathway reveals some key features: (a) leadership actions that invited input, signaled openness to feedback, and communicated a rationale for change that enhanced engagement and motivation of team members, (b) insistence upon an interpersonal climate of psychological safety in which team members felt comfortable to ask questions and speak up about concerns, (c) opportunities for reflective practice, including briefings and debriefings so that team members could more easily understand one another's capabilities and coordinate their actions, and (d) use of data for ongoing review and improvement. These features of the DIEP flap team's work are consistent with features of successful practice change efforts identified by Edmondson, Bohmer, and Pisano (2001).

With shared tacit knowledge of the procedure, team members could better coordinate implicitly (non-verbally), and could better recognize when something was unusual and verbalize (explicate) the situation or concern. By establishing trust and common ground, team members also established an information-rich decision-making environment in which overlapping role and task knowledge were cultivated. This strengthened the team's chances of avoiding errors or inefficiencies due to coordination gaps that were more likely in the prior approach to conducting the DIEP flap procedure. This transition – from individual-centric role performance with differing concepts of the procedure and timing of tasks to conceptual alignment across the team with a high degree of inter-member predictability represents a collective shift of theory in practice, achieved through intentional, joint construction.

Thus far, we have looked at how values in action may appear to veer from the values we espouse at the level of the individual and small group. Next, let us examine how one organization chose to assess whether the safety values espoused at the executive level of a hospital were successfully enacted throughout the organization.

Case Example 6.3: Value Alignment Across Organizational Silos

The following case, from a community hospital, is based on a series of interviews characterized as a 'Vertical Patient Safety Perception Audit.' Individuals in clinical, middle management, and executive roles were asked: 'How do you define safety?', 'Are there any accidents waiting to happen?', 'Does your organization value patient safety?' and 'What is the safety of patient care in your organization on a scale of one to 10, where one is terrible and 10 is excellent?' At the outset of the audit, senior leaders were confident that high scores would be achieved in all of these dimensions, at all levels of the organization.

The consulting team interviewed nursing personnel and physicians in the medical-surgical unit, critical care unit, and other inpatient units. The team also interviewed middle managers in process improvement, quality, and risk management roles. Finally, they interviewed the Chief Executive Officer, Chief Operating Officer, Chief Financial Officer, Chief Medical Officer, and Chief Nursing Officer. It was a vertical slice starting with technicians, nursing personnel and physicians, then middle management personnel, and finally the executive team. The perceptions of clinicians, middle managers, and senior leaders regarding: (a) 'accidents waiting to happen', and (b) overall safety of care in the organization were strikingly different. At least four of the nurses wept as they talked about their concerns for the safety of patients and how much they worried what might happen to the patients they cared for after they went off shift. They were continuously intervening in small, but potentially harmful problems. For example, pharmacy and lab processes were inconsistent and the quality of information flow across shifts and clinical units was highly variable. Sometimes drug allergies were missed, and lab reports on critical tests were often not communicated urgently.

'Accidents waiting to happen' included giving a patient the wrong medication or wrong blood type – there had already been near-misses and hits. Nobody trusted the Electronic Medical Records (EMR) – the information in it was often stale or just plain wrong. Some personnel entered patient care data in the EMR and others kept paper records. Regardless, every individual kept a personal paper record of each patient to try to ensure that they had an accurate record of the care and the care plan. Physicians were frustrated that their patient care orders were often not followed correctly and that blood draws and lab work had not been performed or that lab reports and other critical information couldn't be located. Both nurses and physicians reported that the organization was not responsive to reports of incidents or near-misses – other than to remind clinicians to be careful and pay better attention – or to follow policy. Instead of contending with real problems in the clinical units, there were continual efforts to implement new quality improvement programmes that senior leaders had read about, or seen a presentation on at a conference. They weren't necessarily a bad idea, but they usually did not address the immediate clinical problems that were causing so much distress. The average perception of patient safety in the hospital, on the scale of one (terrible) to 10 (excellent) for 18 frontline clinicians (11 nurses and seven physicians), was a 'two'!

Middle managers seemed more confident than frontline clinicians that patients were receiving safe care, but did discuss recurrent incidents that seemed to stem from problems with communication across shifts and departments of the organization. When incidents happened, whether the patient was harmed or not, the explanation across interviewees was that the cause was usually an error on the part of the clinician. They also mentioned the problem of stale patient information in the EMR, but did not state that they perceived any serious patient safety concerns, or 'accidents waiting to happen'. The greatest concern for quality and risk managers was not about safety, but that the improvement projects they were responsible for were typically conceived by senior leaders and then delegated to them to implement. They had no authority to implement clinical practice changes that these projects typically required. Hence, they were constantly thwarted in their efforts to roll out process improvement programmes, especially those that required physician cooperation, like development of treatment protocols. On average, middle managers (eight) rated the safety of patients in their hospitals at 'eight' – decidedly good.

At the executive level, perception of patient safety was uniformly high, across all roles. The CEO perceived that the hospital was an extremely safe facility in which to receive care, citing the organization's process improvement programmes as evidence of close attention and responsiveness to patient care issues and needs. The executive team (five members) uniformly rated the safety of care in their organization at '10'. The disconnect between the frontline care providers and middle managers regarding perceptions of safety in their hospital was substantial, but the disconnect between the frontline and the executive team was profound.

Although the foregoing example paints a very worrisome picture from a patient safety perspective, this audit actually yielded a very good outcome. Initially shocked by the difference between frontline perceptions of safety and his own, the Chief Executive Officer (CEO) nonetheless embraced the perceptions of clinicians. He revised his theory of the situation, his perception of safety in the hospital – deferring to those who provided hands-on management of the safety and quality of patient care. Resources were allocated to address the major issues identified in the audit and to establish mechanisms for remaining sensitive to frontline safety perceptions and needs, including periodic sampling using the vertical safety perceptions audit. Other organizations have more difficulty responding to frontline problems, as is evident in the following case.

> Recently, our emergency department has received a mandate from hospital administration to lower wait times – the time it takes us to have a patient seen by a physician. Patient satisfaction surveys indicate that over the past month our patients are less satisfied with our care because of high wait times. Our CEO's pay is linked to scores on patient satisfaction surveys, so he pays close attention. We have had quality management working with us to lower wait times, and their response has been to try to streamline triage and improve our 'fast track' process, which gets patients with minor complaints seen more quickly, without using rooms that are needed for patients with higher acuity. They are applying something called 'Lean', which I think is a way of making processes more efficient. This can't hurt, but a big reason we have had higher wait times – that nobody seems to be paying attention to – is because of a new requirement that was imposed on emergency departments throughout our healthcare system. We are just one of about 40 hospitals in our system.
>
> Someone at the corporate level decided to reduce inventory costs by outsourcing orthopedic supplies, like crutches. Nurses and techs are now required to act as a financial agent for the third party vendor of orthopedic supplies. For example, if a patient is to be discharged with crutches, I not only instruct the patient on the use of the crutches, I am responsible for explaining to them that they are responsible for paying a third party vendor for the crutches. I have to walk through the vendor's forms with the patient and get the patient's signature on an agreement to pay. This is often confusing for the patient and I have to explain more than once before they will sign. They don't understand why it isn't just one bill – in fact, patients may not know in advance, but they will usually get at least three separate bills for being seen in the ED, often depending on the diagnostic equipment/processes used, specialty physician services needed, and supplies required.
>
> Anyway, I'd say, on average, acting as a financial agent for the orthopedic supply vendor adds 15 minutes to the discharge process, which delays getting from the waiting room to the exam room to be seen by the doctors. During ski season we get quite a few sprains and breaks, so we really notice the increase in wait times. Serving as a financial agent for a third party vendor has nothing to do with nursing care and should be handled by the finance department or someone with a job like that. Nobody seems to understand that requiring nurses or techs

to do this job bogs down overall performance in the ED. Whatever we save as a healthcare system on inventory must pale next to the ill-will and inefficiency it causes. The doctors, like the CEO, have performance pay linked to patient wait times and they get frustrated with nursing for 'holding them up'. We [nurses] feel a lot of pressure to fix the problem, but can only cut so many corners in our work.

The above example reveals how efficiency measures conceived far away from the frontline environment may cause inefficiency which remains invisible because of a sense of professional obligation to compensate – to 'cut corners'. Unless the organization has sensitive mechanisms in place to detect emergent sources of inefficiency or risk before compensatory adaptations are normalized, it is not likely to have an accurate perception of the 'risk state' of any given unit or department. The Lean Six Sigma initiative mentioned in this case is a business management strategy intended to improve efficiency and quality by identifying and minimizing causes of process variability. While the nurse who shared this case perceived that there may be benefit in applying the techniques of Lean Six Sigma to healthcare processes, he also stated that the disruption to throughput associated with the new nursing task hadn't been noted and wasn't being examined. His perception of what the organization valued was efficiency, cost reduction, and increased volume. This was not consistent with the values printed on various posters throughout the hospital, which were:

- The highest quality and safety of care for our patients.
- Respect for the privacy of our patients.
- A pleasant patient experience of care in our hospital.

Clearly, there was a disconnect between the espoused values and enacted values; consequently, the integrity of the organization's values was being questioned.

Values 'Highlighting': How Organizations Let us Know What is Important

'What we are held to account for strongly influences what we attend to' (Woods 2004). Being held to account for individual fulfillment of tasks, rather than the performance of tasks in close coordination with other care providers, remains a powerful source of fragmentation and patient harm (Patankar, Brown, and Treadwell, 2005; Patterson et al., 2002). Although the extent to which individual versus team accountability varies with the social dynamics of any given unit or organization, the fulfillment of clinical tasks in contemporary hospital-based care is more an individual activity than truly team-based (Dominguez et al., 2005; Gittell et al., 2000; Lawrence, 2002). These conditions have arisen because the social and technological complexity of contemporary healthcare is overwhelming the century-old model of hospital organization and management that was conceived to support the independent practices of physicians (Merry and Crago

2001; Merry, 2005; Sharpe and Faden, 1998; Starr, 1982). Although the current picture may seem bleak, the good news is that clinical environments are rich in diverse intellectual resources which, if harnessed through team processes, may yield unprecedented positive results.

The theory of action perspective can help us map out these possibilities and link them to actions and consequences, but how can we surface and test these educated guesses? How can we use them to influence the direction of safety culture and align safety values and operating assumptions with desired behaviours? Three different models of underlying values are summarized in Table 6.1 that have pervasive influences on behaviour. These values have personal, interpersonal, and organizational influences since they are associated with sources of power, action strategies, psychosocial consequences, and learning outcomes. Model I is based on the value of competition, while Model II is rooted in collaboration (e.g. Argyris, 1977). Model III is centreed on innovation (e.g. Bierly et al. 2000; Wang and Ahmed, 2003). These models present means to classify organizational behaviours based on deeply held values and subsequent transformation of organizational cultures.

The values of Model I may lead to adaptive incremental learning by addressing how to make a current tactic work more effectively to gain competitive advantage. However, the value set underlying this approach is defensive and does not directly confront the reasons behind actions. Moreover, Model I values may result in no learning (e.g. Romme and van Witteloostuijn, 1999) if deception or concealment prevents insight or corrective action. Model II values are more likely to lead to continuous learning improvements since decisions are based on valid information and inquiry into the basis of operating assumptions. Model III makes way for radical change by unlearning ineffective routines, and by inventing new processes to produce novel insights, create knowledge, and drive strategic thinking. Snell and Man-Kuen Chak (1998) describe the change from Model II to III as the difference between a paradigm shift and a paradigm invention.

Underlying values have powerful influences on the prevailing safety culture. Model I values are consistent with a secretive or blame culture; Model II values represent a shift toward a reporting culture; and Model III values support transformation to a just culture.

Acculturation

Typically, acculturation is associated with the change that takes place when two autonomous cultures come into continuous contact with each other. Berry (1983, 2003), Cox (1991), and Nahavandi and Malekzadeh (1988) present conceptual frameworks to understand acculturation as a strategy with four dimensions: assimilation, integration, separation, and deculturation. Further, there are individual-level psychological issues and group-level sociological issues that need to be understood. Both these sets of issues need to be viewed in the context of the broader cultural perspectives and from the vantage point of the two groups rather

	Values	Source of Power	Strategies	Consequences	Learning Outcomes
Model I	Pursue own goals	Advocacy	Control environment for factors of personal relevance	Actor seen as defensive, competitive, manipulative	Self fulfilling process
Model I	Maximize winning	Competition	Own and stay in control of task	Defensive interpersonal and group relationships	Single loop
Model I	Minimize negative feelings	Deception	Unilaterally protect self; blame and stereotype others	Defensive norms: mistrust, rivalry	Attributions untested
Model I	Emphasize rationality	Manipulative Non confrontational	Unilaterally protect others from being hurt; hoard or censor information	Low freedom of choice, internal commitment, and risk taking	Single loop
Model II	Valid information	Openness Combines advocacy and inquiry	Design environments that allow freedom to experience results	Actor seen as minimally defensive, as a collaborator and a facilitator	Processes can be disconfirmed
Model II	Free and informed choice	Trust, respect for others, cooperation	Joint control of task. Seeks high participation	Involvement, commitment, helping behaviours	Double loop
Model II	Internal commitment to decision and personal responsibility	Competence	Self-protection joint with enterprise growth; recognizes inconsistencies and incongruities	Learning oriented norms. Productive confrontation	Public test of attributions
Model II	Monitoring effectiveness of implemented decision	Personal accountability	Bilateral protection of others	High freedom of choice, internal commitment, and risk taking	Double loop
Model III	Creative quality and value innovation	Organizational ambition, wisdom, courage	Competence based creative and strategic thinking	Knowledge creation by radical change, engagement	Triple loop and unlearning to create quantum leaps

Table 6.1 Underlying values, unquestioned assumptions, and strategies influencing performance (adapted from Argyris, 1977; Bierly et al. 2000, Wang and Ahmed 2003)

than that of the researcher. At the individual level, acculturation strategies can also be viewed in terms of attitudinal (what one seeks) and behavioural (what one is able to do). At the group-level, acculturation strategies may be motivated by social survival or advancement to better opportunities (e.g. in ethnic migration) or by business survival or expansion (e.g. corporate mergers and acquisitions). In this section, the emphasis is on acculturation as it relates to corporate mergers and acquisitions. However, the four acculturation strategies mentioned earlier are just as applicable in corporate settings as in social settings. Assimilation refers to complete adoption of the dominant group's identity and values; integration allows for the non-dominant group to maintain its identity and values, but it requires the dominant group to allow the non-dominant group's identity and values to co-exist; separation refers to social segregation of the two groups while allowing for transactional reliance on each other; deculturation is a destructive phenomenon wherein both groups may end up destroying each other in a quest for domination.

In Figure 6.3 we present a conceptual perspective on the degree of acculturation, the four strategies of acculturation, and the impact of each strategy on the social, individual, and corporate aspects. From a corporate perspective, the general objective of most mergers and acquisitions is to gain market share, improve efficiencies, and build a stronger overall organization (Cartwright and Cooper, 1993). However, depending on the acculturation strategy in use, whether intentional or not, the effect of the merger may not be consistent with the plan. Deculturation and separation are viewed as strategies that produce less degree of acculturation, and integration and assimilation are viewed as strategies that

Degree of Acculturation

Minimum ←————————————|————————————→ Maximum

Deculturation	Separation	Integration	Assimilation
Social destruction	Social segregation	Cultural pluralism	Cultural singularity
A+B=>0	A+B=>A+B	A+B=>C	A+B=>A
Individuals unable to cope with stress	Individuals find ways to exist in isolation or within their social/cultural groups; transactional interdependence between A and B	Individuals thrive and enable changes in both organizations	Individuals adopt dominant organization's values
Corporate merger; internal competition and ill-will between employees from organizations A and B result in dysfunctional and self-destructive merger	Corporate merger; individuals and functional units continue to operate as separate units; duplication of functional areas and system	Corporate mergers; integrative values creating "best of both worlds" by cross utilization of people, resources, and technologies; respect for heritage and original values	Corporate acquisition of 'B' by 'A'; 'A' ensures all individuals are imprinted as 'A'; no sign of 'B'; all evidence of B's culture is destroyed

Figure 6.3   Results of various acculturation strategies

produce a higher degree of acculturation. Although assimilation produces most acculturation, it is at the cost of complete dissolution of the non-dominant culture. Many organizational as well as social studies illustrate the value of cultural pluralism and therefore, integration should be viewed as the most desirable acculturation strategy. Nonetheless, depending on the compatibility of values between the merging organizations, integration may not be the best strategy.

Acculturation in Mergers and Acquisitions

Corporate mergers and acquisitions are much more prominent in the aerospace industry than in the healthcare industry. For example, on the manufacturing side, we have seen the mergers of McDonnell Aircraft with Douglas Aircraft to form the McDonnell-Douglas Corporation; which was followed by the merger with the Boeing Company. As evidenced in the corporate name, the McDonnell-Douglas merger appears to be an integrative acculturation, while the McDonnell-Douglas and Boeing merger appears to be an assimilative acculturation. Nonetheless, informal anecdotes from original McDonnell Aircraft employees who experienced the merger with the Douglas Corporation and later witnessed (as retirees) the merger with Boeing recall the following:

> When we merged with the Douglas Aircraft Company, our technologies were different, our tooling was different, our market segments were different and our values were different! We were two independent aerospace companies brought together out of a business opportunity. There were minimal efforts to integrate us from a social or cultural perspective. We didn't get along with the Douglas guys and they didn't care about us. Nonetheless, we were very professional when we had to collaborate. Now, we see similar differences between Boeing and McDonnell-Douglas. The fellows in Seattle (Boeing Commercial Airplane Company) work very differently from us in St. Louis (Boeing Integrated Defense Systems – the old McDonnell Douglas Corporation). Although we have one logo and one company name, we have incompatible tooling, systems, and technologies.

Similarly, many airlines have experienced mergers. For example, Ozark Airlines merged into Trans World Airways, which was later acquired by American Airlines. One pilot who served through both mergers recalls the following:

> I am one of the few original Ozark Airlines pilots. When Ozark was acquired by TWA, I was called a 'green blood' because Ozark Airlines livery was in green and the TWA pilots were called 'red blood' because their livery was in red. Toward the end of my career, we merged with American Airlines, and the history repeated! We were called 'red blood' and the American Airlines pilots were called 'blue blood'. In both instances, when the crews were mixed (e.g. one pilot from TWA and one from American) our interactions were strictly professional –

we never discussed family issues or topics of general interest, which were very common when two pilots from the same heritage were in the cockpit. We always wondered how we would react in the case of an emergency.

The accident of Air Ontario flight 1363 in Dryden, Canada is one of the most illustrative examples of organizational cultures impacting safety (c.f. Maurino, et al. 1997). Among other factors, the investigation revealed that some Air Ontario F28 pilots used the Piedmont F28 Operations Manual while others used the USAir F28 Pilot's Handbook, since Air Ontario did not have its own F28 operations manual. Although both manuals are comprehensive and both obviously deal with the same type of aircraft, there were sufficient differences in the operating procedures of these two carriers to create potential problems on the flight deck. (ICAO Journal, 1995).

Nahavandi and Malekzadeh (1988) present the idea of 'congruence' in the acculturation modes – if the two companies involved in the merger/acquisition process could agree on the specific mode of acculturation, appropriate tactics could be employed to effect a successful merger. For example, if the two organizations agree that the merger was going to occur in accordance with the 'integration' strategy, appropriate steps could be taken to preserve certain aspects of both cultures as well as developing the 'best of both worlds' scenarios. Similarly, if both organizations were to agree on the 'assimilation' strategy, appropriate steps could be taken to imprint the dominant organization's cultural values on the non-dominant organization. Nahavandi and Malekzadeh (1988) further argue that although much of the research on mergers and acquisitions indicates success of related mergers (mergers of companies in same or related industries with opportunities for profit maximization), there is evidence of successful mergers of companies that are not doing business in related industries, unrelated mergers. One common success factor across these two groups is the agreement on acculturation strategy. For unrelated mergers, separation strategy works better than integration or assimilation strategy – the alliance is primarily at the financial level and the operational or technical levels are largely independent.

In an impressive and unique study of 50 mergers and acquisitions, Larsson and Lubatkin (2001), using a case survey method, conclude that the only independent variable producing successful acculturation (they mean 'integration' as defined by Berry (1983)), is social control: 'involve the affected employees in such socialization activities as introduction programs, training, cross visits, joining retreats, celebrations and other such socialization rituals and they are likely to create a joint organizational culture on their own volition, as long as they are allowed autonomy'. Larsson and Lubatkin further claim that if autonomy is not viable, senior management involvement, development of transition teams, and temporary personnel exchange/rotation would bolster the integration success. The significance of socialization tactics is underscored by Cable and Parsons (2001), who discovered that fixed (meaning regular) and sequential socialization tactics were successful in increasing the congruence between employee values and organizational values through shift in employee values. One could argue that meaningful socialization can imprint organizational values on individuals, a form of assimilation. Thus, acculturation strategies are just as significant in mergers and acquisitions as in grooming new employees of a single organization.

Transformation

Again, we are using Rochon's (1998) definition of transformation – it is rooted in organizational values: value connection, value conversion, and value creation. Based on the preceding discussion, value-based transformation can occur in the context of mergers and acquisitions, indoctrination of new employees, as well as redirection of an existing, single organization with its existing employees through the practice of a more compelling set of values. In this section, the Josie King Case is used to illustrate how a value conversion at the individual level – by Josie King's parents – resulted in a change in their strategy regarding how they were going to deal with the tragedy, which in turn resulted in a shift in not only their own attitude toward medical errors, but also that of the healthcare providers interacting with them. Ultimately, Josie King's parents were effective in accomplishing a change in behaviours of not only the healthcare professionals at Johns Hopkins Hospital, but at several hospitals across the nation.

Case Example 6.4: The Josie King Case

In 2001, Josie King, an 18-month-old child, who was being treated for severe burns at Johns Hopkins, died of preventable medical error. This remarkable story is recounted by the child's mother in print (King, 2006, 2009) and in a documentary film (Christopher, 2006). Following the death of her daughter Mrs. King has become a national advocate for patient safety (viz., Josie King Foundation). This case is used to present a comprehensive perspective using the Safety Culture Pyramid.

Safety Performance

Due to a faulty temperature control, one-and-a-half-year-old Josie King was accidentally burned by scalding water in a bathtub at her home. She was rushed to Johns Hopkins Hospital where she initially received excellent care. After about two weeks in the hospital, Josie had made very good progress recovering from the burns and her parents were preparing to bring her home. However, several days before Josie's release things went very wrong at the hospital. The King parents noticed that their daughter was constantly thirsty and crying for something to drink. Josie looked pale and seemed to be losing a significant amount of weight. The parents repeatedly brought these concerns to the attention of the nursing staff and requested that a physician evaluate her. While the nursing staff assured the parents that their daughter was in no danger, Josie experienced cardiac arrest and was placed on life support. Tests revealed that Josie experienced brain death and the parents subsequently removed her from life support. The hospital admitted that Josie had not received proper hydration and that she had wrongly been administered a pain-killer (methadone) which led to her cardiac arrest.

Safety Climate

Johns Hopkins enjoys a reputation as one of the best hospitals in the world. The King parents shared this perception of the hospital and were initially comforted by the fact that their daughter was being treated there. The news of Josie King's death was met with shock and disbelief by the hospital staff. Reports of this tragic medical error caused significant dissonance for them, since they prided themselves on excellent teamwork, communication, and healthcare. It also profoundly called into question the safety of patient care.

A pervasive attitude of the hospital staff's superiority of professional medical opinion was apparent in this case. The King parents reported feelings of frustration that their concerns about Josie's dehydration and deteriorating condition were not being listened to. The parents felt that their opinions were dismissed as overly protective and medically uninformed. The prevailing attitude of trust in the traditional hierarchy of medical opinion about a patient's status was not questioned.

Safety Strategies

The Johns Hopkins Hospital mission is excellence in healthcare. Dr. Dover, the Medical Director, rather than assuming a defensive posture, apologized to the Kings for their daughter's death. The Kings retained a personal injury attorney to represent them. For months, the discussions were adversarial. However, the Kings were eventually able to deal with their emotions of grief, despair, anger, and revenge. They resolved to move in a positive direction to bring some good out of their child's tragic death. The Kings decided to think in a new way. When approached by legal counsel from the hospital with an agreement for a monetary settlement, the Kings decided they didn't want money for their child's death, what they wanted was for Johns Hopkins to change. Remarkably, the Kings sought to collaborate with Johns Hopkins and form a partnership to improve patient safety. The monetary settlement was used to create the Josie King Foundation. Its mission is 'to prevent others from dying or being harmed by medical errors. By uniting healthcare providers and consumers, and funding innovative safety programs, we hope to create a culture of patient safety, together'. One of the Foundation's experts is Dr. Peter Pronovost who was named by the *Time Magazine* as one of the world's 'most influential people' in recognition of his work in patient safety. Sorrel King, Josie's mother, is also widely recognized as passionate advocate for improved patient safety.

Safety Values

The work of Chris Argyris (1977) offers insight into the dynamics of underlying values and assumptions that are seldom questioned. These forces exert powerful influences on decisions and actions in organizational settings. Table 6.1 provides a description of three different models of governing values and their associated sources of power, strategies, consequences, and learning outcomes. Model I (adversarial/defensive values) and Model II (collaborative values) are based on the work of Argyris (e.g. 1977). Model III (innovative values) stems from the work of Bateson (1972), McWhinney (1992), and more recently from Bierly et al. (2000) and Wang and Ahmed (2003). Model III seeks more comprehensive, holistic, and systematic understanding of events by incorporating multiple perspectives to interpret experience. The King case demonstrates movement from Model I (adversarial/defensive values), across Model II (collaborative values) to Model III (innovation values) resulting in transformational learning.

Model I is a very common approach for resolving difficult situations that relies on an adversarial/defensive style. Key values include pursuing one's own goals, maximizing winning, minimizing negative feelings, and emphasizing rationality. Sources of power involve advocacy, competition, deception, and manipulation. In this model common strategies emphasize control, unilateral protection, blaming, stereotyping others, hiding errors, gaming the system, mistrust and rivalry. Many medical errors are approached from a Model I perspective that is further fuelled by anger, grief, fear, and loss. Issues are litigated and the system that led to the errors remains basically unchanged, learning is minimal, and the system is self-sealing.

During negotiations following Josie King's death to settle the lawsuit, remarkably, a new approach eventually developed that demonstrated a shift to Model II values, which emphasizes collaboration. Sorrel King (2006) maintains that families who have experienced medical error are not healed by the monetary settlement. She states that what families really want and need is a sincere apology, the complete truth, and assurance that the problem, which led to the error, has been fixed and will cause no further harm. She argues that if these conditions are present, the pain of the settlement process can be reduced and collaborative efforts can be increased. The Kings heroically managed to redirect their emotional energy from thoughts of revenge for their daughter's death to making a contribution to safety that could prevent errors like this from recurring.

Model II is characterized by collaboration that is based on valid information, free and informed choice, internal commitment and monitoring the effectiveness of implemented decisions. Sources of power for a collaborative approach include openness, inquiry combined with advocacy, trust, respect for others, cooperation, competence, and personal accountability. Strategies for the collaborative model include joint control of the task and high participation from all parties. Consequences of this model are involvement, commitment, helping behaviours, and risk taking. The Kings decided to work in partnership with Johns Hopkins using the settlement money to first establish the Josie King Pediatric Patient

Safety Program in the paediatric intensive care unit. The programme focused on improving communication, teamwork, safety culture, and disseminating best practices to other units.

The Kings' attempt to transform patient safety has matured to express a unique set of values and strategies aimed at fundamental, deep-seated change. It is represented by Model III, which is distinguished by values that embrace creative quality and value innovation. Sources of power include organizational ambition, wisdom, and courage. Strategies involve competence based creative and strategic thinking. Consequences and outcomes of this approach include knowledge creation by radical change, engagement, triple loop learning and the unlearning of ineffective behaviours to create quantum leaps. Now several years later the Josie King Foundation offers a wide array of programmes and resources for 'creating a culture of patient safety, together'. The King couple's quest for improvements did not stop at Hopkins. Some of their partnership's joint accomplishments include: shaping national legislation (the Josie King Act), helping advance Dr. Pronovost's Comprehensive Unit-Based Safety Program (CUSP) and the Patient Safety Group. In partnership with Shadyside and Children's Hospital in Pittsburgh, a Rapid Response Team has been developed that can be called upon by the patient or family. The Rapid Response Team concept was also adopted as part of the 100,000 Lives Campaign by the Institute for Healthcare Improvement. Additionally, the Josie King Foundation offers pilot funding to support the development of innovative patient safety programmes. Sorrel King serves as an indefatigable speaker and champion for the cause of patient safety.

The Kings' accomplishments took courage and driving ambition to challenge assumptions, think in new ways, and approach problems from different perspectives. They have widely disseminated their learning and acted to create a new culture of safety by embracing new strategies and encouraging transformational leadership.

Whole Pyramid Dynamics

Sorrel King states that her daughter's death and the deaths of many others are not merely due to events happening in isolation. It's not just the fault of a single doctor, or the mistake of a single nurse, or only the result of a poorly labelled medication. Medical errors involve systems of interaction between teams of doctors, nurses, and other healthcare workers where miscommunication, misunderstanding, unchallenged assumptions, and the systemic structure of the organization have powerful effects on outcomes.

Conclusion

The King story demonstrates the power of a mutual commitment to reflect on difficult events, to make significant changes, and to widely disseminate the lessons learned. The story also demonstrates how new thinking has resulted in innovative processes and policies being embedded in hospitals throughout the United States to improve the culture of patient safety and save lives. The Josie King case is a powerful story of innovation, transformational learning, wisdom and courage that began with a tragedy.

Chapter Summary

Safety values are the foundation of the Safety Culture Pyramid Model. While personal values are held by individual people, organizational values are those that are shared by the people within the organization. In this chapter, we discussed two methods to discover the enacted values within a group: deep dialogue and narrative analysis. With the help of three case examples, we illustrated how gaps between enacted versus espoused values could be analyzed and how such gaps could be reduced through structured communication and teamwork protocols. Further, the historical context of differences in values among physicians and the hospital staff indicate a transactional relationship between the physicians and the hospital staff, making it difficult to build high-performance teams across professional groups (physicians who are not employees and nurses, therapists, pharmacists, and technicians who are hospital employees). The highly sophisticated and specialized training of physicians, coupled with efficiency expectations, embeds psychological presets that lead to flawed decisions.

Three different models of underlying values were presented. As one moves the organization from Model I to Model II to Model III, there is an associated value conversion from competition, to collaboration, to innovation. Concurrently, there are shifts in sources of power for the people involved, strategies in achieving the goals, consequences associated with the respective strategies, and the final learning outcomes. Ultimately, the transformation of an organization into a Model III organization holds the potential for transformative learning as illustrated in the Josie King case.

Acculturation is a fundamental challenge associated with value creation, conversion, and connection as well as protection. Four acculturation strategies were discussed in the context of corporate mergers and acquisitions and in the context of the self-transformation of a single organization. In classic mergers and acquisitions, the motivation is business-oriented, but the reasons for failures are rooted in cultural incompatibilities. More specifically, mergers and acquisitions fail because the two organizations don't discuss and agree on the acculturation strategy and build the corresponding implementation plan. Research illustrates that both related and unrelated mergers can be successful if appropriate acculturation strategies are employed. Thus, organizational values as well as individual values play a pivotal role in organizational success. From a safety perspective, disconnects in values can lead to confusing procedures, non-communicative attitudes, and ultimately fatal errors.

References

Abma, T. and Widdershoven, G. (2005). Sharing stories: narrative and dialogue in responsive nursing evaluation. *Evaluation and the Health Professions*, 28, 90–109.

Argyris, C. (1977). Double loop learning in organizations. *Harvard Business Review*.

Argyris, C. (1995). Action science and organizational learning. *Journal of Managerial Psychology*, 10(6), 20–26.

Argyris, C. and Schön, D. (1974). *Theory in Practice: Increasing Professional Effectiveness*. San Francisco: Jossey-Bass.

Argyris, C. and Schön, D. (1978). *Organizational Learning: A Theory of Action Perspective*. Reading: Addison Wesley.

Bateson, G. (1972). *Steps to an Ecology of Mind*. New York: Ballantine.

Berry, J. (1983). Acculturation: a comparative analysis of alternative forms. In: R.J. Samuda and S.L. Woods (eds), *Perspectives in Immigrant and Minority Education* (pp. 66–77). Lanham: University Press of America.

Berry, J. (2003). Conceptual approaches to acculturation. In: K. Chun, P. Organista and G. Marín (eds). *Acculturation: Advances in Theory, Measurement, and Applied Research*. Washington, DC.: American Psychological Association.

Bierly, P.E., Kessler, E.H., and Christensen, E.W. (2000). Organizational learning, knowledge and wisdom. *Journal of Organizational Change*, 13(6), 595–618.

Bolman, L. (1980). Aviation accidents and the theory of the situation. In: G.E. Cooper (ed.), *Resource Management on the Flight Deck*. Moffet Field: NASA Ames Research Center.

Brewer, W.F. and Treyens, J.C. (1981). Role of schemata in memory for places. *Cognitive Psychology*, 13, 207–230.

Cable, D. and Parsons, C. (2001). Socialization tactics and person-organization fit. *Personnel Psychology*, 54(1), 1–23.

Cartwright, S. and Cooper, C. (1993). The role of culture compatibility in successful organizational marriage. *The Academy of Management Executive*, 7(2), 57–70.

Christopher, F. (2006). *Remaking American Medicine: Healthcare for the 21st Century*. A PBS Series.

Chun, K., Organista, P. and Marín, G. (2003). (eds). *Acculturation: Advances in Theory, Measurement, and Applied Research*. Washington, DC: American Psychological Association.

Cox, T. (1991). The multicultural organization. *The Executive*, 5(2), 34–47.

Crandall, B., Klein, G., and Hoffman, R. (2006). *Working Minds: A Practitioner's Guide to Cognitive Task Analysis*. Cambridge: MIT Press.

Dominguez, C., Uhlig, P., Brown, J., Gurevich, O., Shumar, W., Stahl, G., Zemel, A., and Zipperer, L. (2005). Studying and supporting collaborative care processes. In: *Proceedings of the Human Factors and Ergonomics Society 49th Annual Meeting*.

Edmondson, A., Bohmer, R. and Pisano, G. (2001). Speeding up team learning. *Harvard Business Review*, 79(9), 125–134.
Gittell, J.H. (2000). Organizing work to support relational coordination. *International Journal of Human Resource Management*, 11(3), 517–539.
Gittell, J.H., Fairfield, K., Bierbaum, B., Jackson, R., Kelly, R., Laskin, R., Lipson, S., Siliski, J., Thornhill, T., and Zuckerman, J. (2000). Impact of relational coordination on quality of care, post-operative pain and functioning, and length of stay: a nine hospital study of surgical patients. *Medical Care*, 38(8), 807–819.
Helmreich, R. and Merritt, A. (1998). *Culture at Work in Aviation and Medicine: National, Organizational and Professional Influences*. Aldershot: Ashgate.
Hofstede, G. (1984). *Culture's Consequences: International Differences in Work-related Values* abridged edition. Beverly Hills: Sage.
Hollnagel, E. (2009). *The ETTO Principle: Efficiency-thoroughness Trade-off*. Farnham: Ashgate.
ICAO (1995). Six years after the Dryden tragedy, many accident investigation authorities have learned its lessons. *ICAO Journal*, 20–25.
King, S. (2006). Our story. *Clinical Pediatric Emergency Medicine*, 7(4), 268–270.
King, S. (2009). *Josie's Story: A Mother's Inspiring Crusade to Make Medical Care Safe*. New York: Atlantic Monthly Press.
Klein, G. (1989). Recognition-primed decisions. In: W.B. Rouse (ed.), *Advances in Man-machine Systems Research* (pp. 47–92). Greenwich: JAI.
Klein, G. (1993). Naturalistic decision-making: implications for design (CSERIAC SOAR 93-1). Ohio.
Larsson, R. and Lubatkin, M. (2001). Achieving acculturation in mergers and acquisition: an international case survey. *Human Relations*, 54(12), 1573–1607.
Lawrence, D. (2002). *From Chaos to Care: The Promise of Team-based Medicine*. Cambridge: Perseus Publishing.
Lee, B., Tobia, A., Yueh, J., Bar-Meir, E., Darrah, L., Guglielmi, C., Wood, E., Carr, J., and Moorman, D. (2008). Design and impact of an intraoperative pathway: a new operating room model for team-based practice. *Journal of American College of Surgeons*, 207(6), 865–873.
Maurino, D., Reason, J., Johnston, N., and Lee, R. (1997). *Beyond Aviation Human Factors*. Aldershot: Ashgate.
McWhinney, W. (1992). *Paths of Change: Strategic Choices for Organizations and Society*. Newbury Park: Sage.
Merry, M. (2005). Hospital-medical staff culture clash: is it inevitable or preventable? The challenge of governance. The Healthcare Trustees of New York State. Retrieved from <www.dynamichs.org/articles/2-Challenge-of-Gov-MMerry-May2005.pdf> on May 1, 2010.

Merry, M. and Brown, J. (2001). From a culture of safety to a culture of excellence: quality science, human factors, and the future of healthcare quality. *Journal of Innovative Management*, 7(2), 29–46.

Merry, M. and Crago, M. (2001). The past, present and future of healthcare quality. *The Physician Executive*, September–October 2001.

Meyer, J. (2003). Tell me a story: eliciting organizational values from narratives. *Communication Quarterly*, 43(2), 210–224.

Nahavandi, A. and Malekzadeh, A. (1988). Acculturation in mergers and acquisitions. *The Academy of Management Review*, 13(1), 79–90.

Patankar, M.S. and Taylor, J.C. (2004). *Risk Management and Error Reduction in Aviation Maintenance*. Aldershot: Ashgate.

Patankar, M.S., Brown, J.P., and Treadwell, M. (2005). *Safety Ethics: Cases from Aviation, Medicine, and Environmental and Occupational Health*. Aldershot: Ashgate.

Patterson, E., Render, M., and Ebright, P. (2002). Repeating human performance themes in five healthcare adverse events. In: *Proceedings of the Human Factors and Ergonomics Society 46th Annual Meeting*. Santa Monica: California.

Rochon, T.R. (1998). *Culture Moves: Ideas, Activism, and Changing Values*. Princeton: Princeton University Press.

Romme, A. and van Witteloostuijn, A. (1999). Circular organizing and triple loop learning. *Journal of Organizational Change Management*, 12(5), 439–454.

Sharpe, V. and Faden, A. (1998). *Medical Harm: Historical, Conceptual, and Ethical Dimensions of Iatrogenic Illness*. Cambridge: Cambridge University Press.

Snell, R. and Man-Kuen Chak, A. (1998). The learning organization: learning and empowerment for whom? *Management Learning*, 29, 337–64.

Starr, P. (1982). *The Social Transformation of American Medicine*. Basic Books.

Taylor, J. and Patankar, M.S. (2000). The role of communication in reduction of human error. In: Proceedings of The 14th Annual FAA/CAA/Transport Canada Human Factors in Aviation Maintenance Symposium, [CD-ROM: 15.0/1-27]. Vancouver, B.C.

Tversky, A. and Kahneman, D. (1973). Availability: a heuristic for judging frequency and probability. *Cognitive Psychology*, 5.

Tversky, A. and Kahneman, D. (1974). Judgment under uncertainty: heuristics and biases. *Science*, 185, 1124–1131.

Wang, C. and Ahmed, P. (2003). Organisational learning: a critical review. *The Learning Organization*, (10)1, 8–17.

Woods, D.D. (2004). Conflicts between learning and accountability in patient safety. *De Paul Law Review*, 54, 485–502.

Chapter 7
Safety Culture Transformation

Introduction

In the previous chapters, we discussed the Safety Culture Pyramid, starting from the tip, Safety Performance, to the base, Safety Values. This Pyramid helps us describe how safety performance, climate, strategies and values are aligned to produce the temporal state of safety culture. Now, we turn our attention to shifting that dynamic balance toward a more desirable state of safety culture. Previously, we have discussed various *states* of safety culture along the accountability scale and along the learning scale. We have also used Rochon's (1998) definition of transformation. Therefore, for an organization to transform, rather than change, the state of its safety culture from say blame culture to just culture, it will need to start with re-examination of its values, build strategies that are consistent with the newly stated or recommitted values, periodically measure the safety climate to assess its consistency with the intended goals, and then monitor the safety performance to ensure that behaviours consistent with the just culture are rewarded.

In a study of acculturational and anthropological theory, Voget (1963) noted that culture is a state of dynamic equilibrium achieved by a range of individual variables, their relationships with one another, their individual and collective values and motivations, and their relationships with their environment. As such, 'change begins with alterations in the kind, rate, and intensity of interactions that link individuals to the significant institutionalized patterns of the system'. Therefore, in order to effect a change from blame culture to just culture, one has to change the kind, rate, and intensity of the interactions among individuals such that they are consistent with those espoused in a just culture. Also, according to Voget, for a change to be considered *cultural*, it has to be irreversible.

The Need for Cultural Change

Mitroussi (2003) argues that when the survival of the organization is at stake, its people will be more willing to give up old values and practices and take up new ones. So, an organization might be amenable to a change in its safety culture if there is an appropriately intense and urgent internal or external threat. In the case of aviation, the International Civil Aviation Organization's (ICAO) requirement to implement a Safety Management System and improve safety culture might serve as a sufficient external threat. However, it is also well recognized in the aviation industry that safety is a business and operational necessity. Although the specific business benefits of discrete safety programmes have not been fully established, the impact of catastrophic accidents on the overall financial health of an airline is clear (e.g. Eastern Airlines, Pan Am, TWA, ValuJet, etc. have all failed soon after catastrophic accidents). The healthcare industry, on the other hand, is making steady progress toward reducing preventable injuries and deaths through organized efforts – the National Patient Safety Goals are serving as common goals for the industry. So, in both industries, the need for a cultural, therefore irreversible, change is widely acknowledged.

A Method for Cultural Change

Most of the safety culture studies have focused on the *concept* of safety culture. 'Exactly how to create a safety culture is not clear, although many agree that it will include continuous organizational learning from "near miss" incidents as well as accidents' (Ringstad and Szameitat, 2000). Theoretically, cultural change efforts may start from the bottom up or from the top down. In safety culture transformation, in particular, there is a need for both bottom-up and top-down alignment. While leaders need to establish policies and provide resources to support the policies, all the employees need to participate in the actual transformation effort. Knowles (2002) demonstrates the significance of the top leader in the organization stepping out of his/her comfort zone, facing the safety facts, and sincerely pledging to make the necessary changes.

The cultural change direction pursued by the top management needs to be translated for middle managers because they play a critical role in this cultural transformation processes. Guth and MacMillan (1986) claim that the degree to which middle managers support the new initiatives will determine the rate at which the initiatives will be successful or if they will be successful at all. Applying Guth and MacMillan's perspective to a specific example, if the top leaders are committed to moving their organization from a blame culture to a just culture by introducing a non-punitive error reporting system, the middle managers need to: (a) agree that changing the organizational culture is the right corporate objective, (b) agree that implementation of a non-punitive error reporting system is the right strategy, and (c) see that the corporate objective is consistent with their individual self-interest.

Hendrich et al. (2007) report the journey of Ascension Health toward a corporate goal of 'zero preventable deaths and injuries'. They present the following key steps in accomplishing their cultural change: 'Establishing a sense of urgency, creating a guiding coalition, and developing the Destination Statement II'. The Destination Statement II most vividly describes the desired safety culture. While all of the steps involved in achieving the transformation – and characterizing the transformation as an ongoing journey rather than a destination – are important, building a coalition to rally in support of the desired culture is consistent with findings from other domains (Guth and MacMillan, 1986; Fernández-Muñiz, Montes-Peón and Vázquez-Ordás, 2007).

Patankar and Sabin (2008) describe a safety culture transformation effort in the Technical Operations domain of the Air Traffic Organization of the Federal Aviation Administration. Based on the pre-post interventions surveys, they established that a non-punitive error reporting system could be used to shift away from the predominantly blame-oriented safety culture. This study also demonstrated the importance of longitudinal studies that use a quasi-experimental approach to illustrate the effects of an intervention.

How Long Will it Take to Change the Safety Culture?

The question of how long it would take to change a given culture is common among managers, but the response depends on the scale (how many people need to change their behaviour?) at which such a change is desired. Based on Moore's (1991) classification, it will depend on the relative proportion of early adopters, sceptics, and critics of the change programme. Also, there is a notion of 'organizational readiness' for safety culture change – described in terms of preconditions such as the levels of compliance with standard operating procedures, collective commitment to safety, individual sense of responsibility toward safety, and employee-management relationship (Patankar, 2003). Literature on fostering cultural change suggests that focused and deliberate efforts are essential to craft such a change (Sabin, 2005).

Additionally, there are leverage factors such as level of awareness in the greater community, regulatory requirements/pressures, business survival factors, industry standards, etc. When local change efforts transition toward organization-wide changes and mature into a policy/regulatory change, the change programme tends to achieve a much higher degree of stability. For example, if we trace the evolution of the Crew Resource Management (CRM) Program (Wiener, Kanki, and Helmreich, 1993) and the Maintenance Resource Management (MRM) Program (Taylor and Patankar, 2001), one glaring difference as to why the CRM programme has been institutionalized and the MRM programme has not, is that CRM training is a regulatory requirement. Also, the global awareness and acceptance of the CRM programme was greatly enhanced by practitioner heroes such as Captain Al Haynes, multiple airline pilot unions, and the operational leaders of many airlines who are pilots. The pilots had their own share of challenges in overcoming the deep-rooted culture of command and control, but large-scale efforts leveraged by the above mentioned factors have led to a time when most novice pilots are familiar with the key concepts of CRM and expect their future airline jobs to demand refined practice of CRM techniques (Patankar, Block, and Kelly, 2009).

In contrast, the maintenance community suffered significantly from not having the MRM training as a regulatory requirement. As the financial strength of most US airlines declined in the 'post 9-11' era, MRM programmes were disbanded at many of the airlines and champions of such programmes lost their jobs. Well-recognized spokespersons like past National Transportation Safety Board Member John Goglia continued to promote MRM programmes, and with the new group of maintenance personnel in charge of safety, many of the companies have turned their attention to the maintenance version of the Aviation Safety Action Program (ASAP). This is an FAA-endorsed programme and has demonstrated its effectiveness in effecting specific changes at both organization and industry levels. Therefore, it seems as though the maintenance industry is set to 'leapfrog' and build the ASAP programmes and incorporate the MRM concepts in the comprehensive solutions that are generated in the ASAP recommendations (Patankar, Block, and Kelly, 2009).

Best Practices in Safety Culture Transformation

Based on the literature on cultural transformation, we offer the following essential steps:

- Demonstrate the need for a cultural transformation.
- Build a guiding coalition to support the transformation.
- Develop a Destination Statement (what will the world look like after the transformation and what will the people entrenched in the current culture do in the new culture? Visualize the destination from the people's perspective).
- Establish specific performance goals with a specific timeline.
- Use assessment tools to objectively measure the baseline performance and the effects of interventions.
- Change individual and team evaluation and reward metrics to align with the desired behaviours.
- Provide training, equipment, and human resources to facilitate the change.
- Publicize success, present awards, and celebrate key milestones.
- Constantly express faith in people's ability to build a better safety culture.
- Remove non-productive individuals.

Step 1: Demonstrate the Need for a Cultural Transformation

First, the process is 'transformational' because it involves change in values – value conversion, value creation, and value connection (Rochon, 1998). In order to achieve such a change, one has to demonstrate why previously (or currently) practiced values are no longer acceptable. In aviation, if we continue to operate under blame culture (value → punishment) we will not be able to address latent systemic failures and will not be able to bring down our accident rate. If the accident rate remains the same and the traffic volume increases, the frequency of aircraft accidents will increase. One can also argue that accident frequency and traffic volume are dynamically associated such that if the accident frequency increases, the traffic volume will drop; essentially, the accident rate will serve as the governor of traffic growth. Most of the technical and crew communication errors have been removed from the system; now, the next level of errors reside deep in organizational safety culture. The biggest challenge to our values is profit versus safety; production versus protection; speed versus accuracy; or efficiency versus thoroughness. Since deregulation, the airline industry has been operating with a close eye on the profit margin – perhaps, with the philosophy of 'no margin – no mission' because the profit margins have been so thin, if any, and so many airlines have been in and out of bankruptcy protection, the overbearing thought among some top executives has been, 'let's see if we make it through this quarter'. Now, it's time to shift the philosophy to, 'no mission – no margin'. Of course, programmes like the Safety Management System will become a mission as well as a business necessity because, in the future, airlines will not be able to operate without a robust safety management system; so, the fixed costs of an airline will increase and further stress the efficiency mandate.

Given these environmental pressures, how can leaders make the case for transforming safety culture? In the FAA Tech Ops case discussed earlier (and presented in detail in the next chapter), senior leaders made it clear to the employees that the agency's funding formula was at risk and change would be necessary to maintain safety and viability (both of the agency and of commercial airlines). In the preceding era, the unspoken value was 'safety first, at any cost'. The new rule-of-thumb was 'safety first – but you need to watch the costs as well'. In other words, we will no longer support safety at ANY cost; we need to balance efficiency and thoroughness. This change struck many in Tech Ops as a basic change of mission; they could not imagine that new systems, structures, and processes could provide the 'backup' which had always been supplied by dedicated human effort in times of trouble and by redundant staffing in normal times. By persisting with the new mandate, senior leadership gained enough credibility to implement a major reorganization, streamline workflow, and implement new information technologies. And while the culture did begin to shift from Blame and Reporting, these initiatives did not create enough 'value conversions' to result in commitment to the new values throughout the organization. At the time of writing, safety culture transformation (i.e. to Reporting and Just Culture as a matter of course) remains a work-in-progress in the FAA (based on promising but still evolving efforts, e.g. ASAP and CRM programmes in the Air Traffic Organization).

In the healthcare environment, making the case for safety culture transformation usually requires translating it into benefits for front-line workers. We know that medical errors are very costly and have been the cause of major national reports (e.g. *To Err is Human*, *Crossing the Quality Chasm*), campaigns such as the 100,000 Lives Campaign of the Institute for Healthcare Improvement, and 'No Preventable Error' initiatives by organizations such as Ascension Health and Beth Israel Deaconess. Yet, changes in practice and transformations in culture usually come from more ground-level approaches like the one presented in the DIEP Flap case in Chapter 6. There, the motivation to change was less about lofty safety goals and more about the felt instinct of the team, as articulated by the surgeon, 'lots of things about this are not working right; we need to fix them to make everybody's life better, including that of the patient'. The practical benefits for front-line staff, as well as the values connection to the mission and vision of the hospital and external safety mandates, provided the compelling case for change in this instance.

Steps 2–4: Build a Guiding Coalition to Support the Transformation; Develop a Destination Statement; and Establish Specific Performance Goals with a Specific Timeline

After making the case for safety culture transformation, leaders are faced with a cluster of necessary elements to make the change happen. The first of these is building a coalition for change. In the case of the DIEP Flap case, this started with Don Moorman, but widened to include key nursing and physician leaders. At North York General Hospital, the change was sponsored by the CEO (Bonnie Adamson), led by the Senior Vice-President of Quality (Susan Kwolek), and facilitated by internal and external consultants (the Quality, Safety, and Risk group at the hospital, with help from Quantum Transformation Technologies as the external consulting partner). With the FAA Tech Ops situation, the change had executive sponsorship by the Chief Operating Officer of the entire Air Traffic Operations and the Senior Vice-President of Tech Ops, but was carried out by a coalition of operations managers with help from the Saint Louis University safety culture researchers. In each case, having a small but committed cadre form a guiding coalition was a key success factor in transforming safety culture.

Developing a destination statement is a well-known feature of successful change efforts (Senge, 1990; Kaplan and Norton, 1998). We see evidence of this throughout the cases reviewed in this book, including the DIEP Flap case discussed previously and the North York General case and the FAA Tech Ops initiative presented in the next chapter. Of equal significance is the impact when such a statement is missing or unbalanced. For example, take the case of ValuJet flight 592, which crashed in 1996. The NTSB investigation of this crash indicated that the airline's operating norms, values, and practices were so heavily tilted toward efficiency that there was literally no room for effectiveness, thoroughness, or quality in the organization's mission, vision, or destination statement. Production values were so dominant that safety was no longer in the picture. According to the NTSB, the crash was caused by a fire in the forward compartment of the aircraft which resulted from improperly stored and sealed reserve oxygen tanks. But the investigation revealed that, behind the scenes, this was an 'accident-waiting-to-happen'. Safety procedures had been routinely ignored, policies had not been followed, and the airline's key priority had been to reduce costs and increase production and throughput. Employees got the message; they cut corners until the point when a disaster happened. Meanwhile, the NTSB also uncovered systemic risk that had gone undetected. Worried about the ValuJet operation, the local FAA inspector had called for more staff to help with monitoring the airline, but was overruled by regional and national FAA staff. In brief, the implied destination statement of ValuJet ('value production at all cost, including that of safety') was noticed by both employees and FAA inspectors but was not rebalanced before it was too late.

Finally, successful efforts to transform safety culture need to link the effort to performance targets and monitor them over time. The campaigns of Ascension Health and Beth Israel Deaconess ('no preventable errors') are good examples of this. So was IHI's 100,000 Lives Campaign. As yet, though, we are not aware of efforts to link underlying changes in Safety Culture (e.g. from Blame or Reporting to Just) to 'hard' data on performance. Thus far, the closest efforts of this kind rely on survey data as proxies for changes in underlying values and culture (again, see FAA Tech Ops and North York General Hospital cases in the next chapter). There is emerging work, however, to suggest that changes in safety culture can be linked (both directly and indirectly) to improvements in 'hard' performance metrics. For example, consider the following points:

- The improvements in patient flow at North York General Hospital, while apparently attributable to Lean Kaizen methodology, can also be linked to the prior culture shift journey; the movement from Blame toward Reporting and Just Culture, and the recognition of shared accountability for patient flow that emerged from the story-mapping session previously discussed.
- Martin Merry suggests that the 'theoretical limit' of a First Curve system is a Four Sigma error rate, while the same limit in a Second Curve system is Six Sigma. These metrics appear to correlate closely with the movement in an organization or system from Blame, to Reporting, to Just Culture (Merry, 2003).
- Amalberti et al. (2005) indicate that a Six Sigma performance level is possible in all of healthcare, though they assert that a performance metric of seven to nine Sigma (previously attained by commercial aviation and some parts of the chemical and nuclear industries) is attainable in healthcare only by relatively stable areas of practice. Nonetheless, their argument indicates at least a correlation between the level of safety culture maturity in an organization and the potential and actual performance in safety of that same organization.

We are seeing the beginnings of a base of knowledge and experience that can help us establish specific performance goals, timelines, and linkages between safety culture maturity and 'hard' data on error rates and other system-level performance metrics.

Steps 5–7: Use Baseline and Recurrent Assessment Tools to Objectively Measure the Effects of Interventions; Change Individual and Team Evaluation and Reward Metrics to Align with the Desired Behaviours; and Provide Training, Equipment, and Human Resources to Facilitate the Change

This section of the change model begins to form the reinforcing practices, structures, metrics, and people strategies needed for an organization to internalize the desired change and build it in to daily routines, norms, and expectations. Up to this point, the change will have been embraced by early adopters (Moore, 1991) but the organization will not have crossed the 'chasm' to the pragmatic majority. Once the change is acknowledged and supported by the formal evaluation system, it will be adopted by those who previously felt it was a 'fad' or that the organization was not serious about implementing it over time. By using ongoing measurement, training, and performance reviews, an organization can signal that it is serious and that the new approach is not going away. In addition, by this time the change will have shown early wins and will have gained commitment from a critical mass of early adopters. This will make it more difficult for the pragmatists to ignore or challenge it.

For example, Aviation Safety Action programmes (ASAP) have been enthusiastically adopted by proponents in commercial aviation and the FAA, including flight, maintenance, and air traffic control. Advocates of the programme view it as self-evident that many benefits, including improved safety, will result from reporting errors, investigating them, deriving key lessons, and disseminating lessons more broadly. This belief has gained momentum in many places in commercial aviation. Yet in recent years, lawyers in potential lawsuits arising from crash have challenged the confidentiality/protection of ASAP programmes, which is one of the basic premises of voluntary, non-punitive reporting as promised to participants in the programme. Adherents of the programme contested this change and the courts did not support them, but they value the programme so much that they continue to support it even when there are legal limits to the protection offered to the reporters.

It can be difficult to 'objectively measure the effects of interventions' when one is trying to measure 'transformation.' Recall that transformation entails changes in values, such as value conversion, value creation, and values connection. But how can one measure such changes over time? In our view, various evaluation methodologies can be used, including surveys and critical incident review as well as qualitative and quantitative analyses of safety strategies, myths, legends and heroes, and underlying values. The choice of methods will depend on the researchers' interest in espoused values, values-in-practice, or both. More dynamic approaches to evaluation involve dialogue, guided inquiry, and storytelling. This is because changes in values often show up as changes of interpretation and sense-making before they can be inferred from surveys or other self-report tools. For example, as part of the Tech Ops case, one of the authors coached a senior operations executive on how to 'learn from disconnect' and help his team to do the same. Specifically, during a time of transition the executive thought the team had agreed to align and coordinate their communications with direct reports about changes in jobs, levels and locations during a time of downsizing and streamlining. Later, he discovered that each area – Western, Central, and Eastern – was handling its communication differently, and that interpretations of the company's direction and intent varied widely across the three service areas. The executive saw this as a 'disconnect' and as a violation of trust; in effect, he felt the team had violated two of their espoused values: 'move together', and 'talk about it before acting if you disagree with the team's direction'. In order to repair the disconnect and create new, shared values, the executive worked with us to apply learning organization tools (surfacing dilemmas, balancing advocacy and inquiry) to the problem, bring it up with his team, and revisit both their values-in-action and their espoused values. In addition, at our suggestion, he called together his extended management team (including the top three layers of management) to share their 'stories' and observations of the change-in-progress, form a shared understanding of where they were and where they were going, and (in the process) enact Rochon's process of 'values conversion, values creation, and value connection'. In effect, he used stories and dialogue with his managers, and inquired into their observations to augment ongoing employee surveys. The use of these additional methods enabled him both to 'measure' the organization's transformation-in-progress and to reinforce desired values and behaviours at a time when the change was new enough that the organization was not ready to implement changes in structures and incentives.

Lastly, we have found that providing training, equipment, and human resources can help to sustain transformational change. North York General Hospital invested heavily in Lean training, both to equip staff with the methods and also to embed the methods in the culture by hiring a 'black belt' in Lean Six Sigma as well as six improvement coaches who report to him. This investment in staff was designed to support the successes the organization has had with Lean and to bring the skills and capabilities in-house. Team-based practice must be reinforced by ongoing training in areas such as Crew Resource Management and cross-discipline simulation. And such training cannot be simply focused on technique; it must get at the values, leadership, and resilience required by a well-performing team. The use of debriefing and clinical process improvement in the DIEP Flap case is one example of this. Another is the use of on-site simulation to prompt cross-unit conversation about learning and core values. Working with high-reliability teams at a major health system in Illinois, two of our colleagues ran an on-site simulation of a trauma case to engage the entire 'team' of the ER, the OR, and the Trauma Bay in observing and reflecting on their values-in-action during the simulation. The research team videotaped the entire simulation, using an actor-patient (a dummy), and they wired microphones for the lead nurses and doctors help record the communication regarding changes in the patient's condition. Then, the team debriefed the videotape, and the experience, with all 80 participants in an auditorium. A key standout was the divergent values-in-action manifested by Trauma and the OR with respect to 'patient stability' at the moment of handoff between the two areas. Trauma intended to convey that patient was destabilizing, which is a cause for urgency and growing concern. The OR physician lead heard the communication as one of concern, but not of urgency, due to different interpretations of blood pressure and other vital signs relative to 'stability' of the patient. The lack of specificity in the Trauma physician's interpretation, combined with the OR physician's interpretation of the communication that the patient was not in the danger zone, could have been a serious patient safety and clinical problem in a real situation of this kind. In short, this form of training reinforced the need for high-reliability teaming between the ER, Trauma, and OR, and fuelled the commitment of the entire staff to persisting in transformational, values-based change.

Steps 8–10: Publicize Success, Present Awards, and Celebrate Key Milestones; Constantly Express Faith in People's Ability to Build a Better Safety Culture; Remove Non-productive Individuals.

Whether a person is fan of Skinner's (1976) operant conditioning, Wheatley's (1999) living organizations, or the Pygmalion Effect (Livingston, 1988), the bottom line is: reward and penalty systems form the rails of the safety culture transformation efforts. On the reward side, positive achievements need to be communicated promptly, success must be celebrated openly, and heroes must be recognized publicly because these acts symbolize the new values that will be supported in the organization. In parallel, behaviours that are inconsistent with the new values need to be penalized, and when necessary, non-productive individuals need to be removed. Management actions as well as inactions send signals to the employees, and the employees (or contractors) learn to read these signals and adjust their behaviours. In knowledge-intensive industries like aviation and healthcare, people take great professional pride in their work; therefore, regular and genuine appreciation of their work is possibly the strongest motivator.

While Skinner's (1976) theories are based on reward and punishment to shape certain behaviours, and Wheatley's (1999) theory of living organizations focuses on the natural tendency of an organization to adapt to survive, both theories engage the reward-punishment paradigm to influence individual or group behaviours. The Pygmalion Effect (Livingston, 1988), on the other hand, speaks to the importance of a manager's belief in a group or an individual's ability to succeed – such a belief is a significant motivator, besides any rewards and penalties that may exist.

Chapter Summary

Safety culture transformation is about achieving a value-based shift in the culture from the existing state to a more desirable state. An audit of current values – at the individual and organizational levels, as well as espoused versus enacted – will help managers develop transformation strategies. Once the desired values, and gaps between espoused versus enacted values, are identified, appropriate changes will have to be made in structures, processes, and policies, particularly in performance expectations and reward mechanisms. Regular measurement of safety climate will provide a longitudinal assessment of attitudes and opinions regarding safety, serving as one measure of whether or not the interventions are producing the desired effects. Finally, behaviours or safety performance will have to be specified and monitored on a regular basis and not just in response to undesirable outcomes. In fact, undesirable behaviours need to be addressed regardless of the outcomes. Recognition and recognition systems should focus on developing the key desired behaviours; penalty systems should focus on curbing the undesirable behaviours, regardless of the outcomes; and general motivation and communication systems should focus on positive reinforcement – believing in the ability of the team to achieve success.

References

Amalberti, R., Auroy, Y., Berwick, D., and Barach, P. (2005). Five system barriers to achieving ultrasafe health care. *Annals of Internal Medicine*, 142, 756–764.

Fernández-Muñiz, B., Montes-Peón, J., and Vázquez-Ordás, C. (2007). Safety culture: analysis of the causal relationships between its key dimensions. *Journal of Safety Research*, 38, 627–641.

Guth, W. and MacMillan, I. (1986). Strategy implementation versus middle management self-interest. *Strategic Management Journal*, July/August, 313–327.

Hendrich, A., Tersigni, A., Jeffcoat, S., Barnett, C., Brideau, L., and Pryor, D. (2007). The ascension health journey to zero: lessons learned and leadership perspective. *The Joint Commission Journal on Quality and Patient Safety*, 33(12), 739–749.

Kaplan, R. and Norton, D. (1998). The balanced scorecard – measures that drive performance. In: *Harvard Business Review on Measuring Corporate Performance*, 123–146. Boston: Harvard Business School Press.

Knowles, R.N. (2002). *The Leadership Dance: Pathways to Extraordinary Organizational Effectiveness*. Niagara Falls: The Center for Self-Organizing Leadership.

Livingston, J.S. (1988). Pygmalion in management: a manager's expectations are the key to a subordinates performance and development. *Harvard Business Review*, September–October.

Merry, M. (2003). Healthcare's need for revolutionary change. *Quality Progress*, 36(9), 31–36.

Mitroussi, K. (2003). The evolution of safety culture of IMO: a case of organizational culture change. *Disaster Prevention and Management*, 12(1), 16–23.

Moore, G. (1991). *Crossing the Chasm: Marketing and Selling High-tech Products to Mainstream Customers*. New York: HarperCollins.

Patankar, M.S. (2003). A study of safety culture at an aviation organization. *International Journal of Applied Aviation Studies*, 3(2), 243–258.

Patankar, M.S., Block, E., and Kelly, T. (2009). Integrated systems approach to improving safety culture. Presented at the Maintenance Aviation Safety Action Program Information Sharing Meeting, Dallas, March 5, 2009.

Patankar, M.S. and Sabin, E. (2008). Safety culture transformation project report. Report prepared for the Federal Aviation Administration. Available from <http://hf.faa.gov>.

Ringstad, A.J. and Szameitat, S.A. (2000). Comparative study of accident and near miss reporting systems in the German nuclear industry and the Norwegian offshore industry. In: *Proceedings of the Human Factors and Ergonomics Society 44th annual meeting* (pp. 380–384).

Rochon, T.R. (1998). *Culture Moves: Ideas, Activism, and Changing Values*. Princeton: Princeton University Press.

Sabin, E. (2005). Promoting the transfer of innovations. In: M.S. Patankar and J. Ma (eds), *Proceedings of the Second Safety Across High-Consequence Industries Conference*, Volume 1: St. Louis: Parks College of Engineering, Aviation and Technology.

Senge, P. (1990). *The Fifth Discipline: The Art and Practice of the Learning Organization*. New York: Doubleday.

Skinner, B.F. (1976). *About Behaviorism*. New York: Random House.

Taylor, J. and Patankar, M.S. (2001). Four generations of MRM: evolution of human error management programs in the United States. *Journal of Air Transportation World Wide*, 6(2), 3–32.

Voget, F. (1963). Cultural change. *Biennial Review of Anthropology*, 3, 228–275.

Wheatley, M.J. (1999). *Leadership and the New Science: Discovering Order in a Chaotic World* (second edition). San Francisco: Berrett-Koehler Publishers.

Wiener, E.L., Kanki, B.G., and Helmreich, R.L. (1993). *Cockpit Resource Management*. San Diego: Academic Press.

Chapter 8
Conclusion

Introduction

Safety Culture is presented in the form of a Safety Culture Pyramid Model, which is a four-layer model with safety values at the base, followed by safety strategies, then safety climate, and finally safety performance at the top. We have presented literature to support our assertion that values influence strategies, both values and strategies influence climate, and ultimately, values, strategies, and climate influence behaviours. Further, we assert that safety culture may exist in a variety of states, which in turn may be presented along a continuum. For example, the accountability scale is used to present the states of secretive culture, blame culture, reporting culture, and just culture. Similarly, the learning scale is used to present the states of failure to learn, intermittent learning, continuous learning, and transformational learning.

A variety of assessment tools and techniques need to be used to fully describe safety culture at a given organization. A simple safety climate survey is not sufficient. Safety performance is usually assessed through case analysis, and root cause analysis is embedded in this approach to determine the contributory causes. Safety performance can also be tracked in terms of specific goals associated with reduction of accidents or incidents, particularly those that are tied with national or international goals. Such accidents/incidents are deemed to have contributory factors that can be tracked and managed to drive down the probability of failure. In such an approach, the causal factors identified at the top of the Safety Culture Pyramid are typically associated with appropriate intervention strategies at the third layer of the Pyramid, with limited attention to Safety Climate or Safety Values, the other two layers of the Pyramid.

Safety Climate is usually assessed with the help of a survey. This is perhaps the most assessed layer of the Safety Culture Pyramid; unfortunately, safety climate measurement has also served as a proxy for safety culture. Most studies focus on one-time measurement of safety climate and no diagnostic advice or interventions to improve the safety climate are suggested. In this chapter, we present two comprehensive case examples: one from the Air Traffic Organization's Technical Operations and one from New York General Hospital. Both these examples illustrate how specific interventions may be developed and safety climate may be improved.

Safety Strategies include comprehensive system-wide strategies like the Safety Management System as well as smaller scale strategies that can be employed at organizational level. Regardless of the scale of implementation, the strategies should have measurable safety performance goals, opportunities for local customization by the various functional groups in the industry, commitment from the top management that appropriate resources will be provided to enable the change recommendations that will result from the implementation of the strategies, and a commitment to change the evaluation and reward systems to align with the stated safety performance goals.

Safety Values and Unquestioned Assumptions are perhaps the most challenging to uncover because they require deep dialogue, which in turn requires a certain level of trust between the interviewer and the interviewee. Story analysis or narrative analysis techniques can be used to get people across the organization talking about their safety-related experiences and these narratives can then be analyzed for common themes, which tend to represent the enacted values in the organization. These enacted values can then be compared with the espoused values, which are typically available in marketing materials or annual reports. The tighter the congruence between the espoused and the enacted values, the stronger the safety culture.

Assessment of Safety Culture

While safety culture needs to be assessed with a variety of tools and methods, for it to be proactive it should start with the assessment of value congruence. Both deep dialogue and narrative analysis could be used as effective techniques to elicit the degree of value congruence. Additionally, people who personify the espoused values could be identified to serve as both role models and key informants to interview in detail. A variety of examples are presented in this book to illustrate how heroes can be found at all levels of an organization, and sometimes even outside the organization.

A study of acculturation at the subject organization would be very helpful in understanding how key values are imprinted on new employees as well as on members of a non-dominant organizational unit by a dominant unit. Since leaders are instrumental in imprinting the desired values on the organization, detailed interviews with top management, including Board members, will provide valuable insight into the values that are being, or are likely to be, imprinted on the organization.

Once the value congruence study is complete, artefacts could be collected from a number of organizational units to test the 'visibility' of those values. Examples of artefacts include, but are not limited to, the following:

- Annual reports.
- Mission, vision, and value statements.
- Organizational performance goals.
- Management performance goals.
- Employee and management evaluation tools (blank forms).

- Employee and management compensation/reward systems.
- Key employee recognition awards and their criteria.
- Organizational charts with roles and responsibilities (determine who is responsible for safety performance and what level of access that position has to the CEO).
- Evidence of organizational commitment to safety performance (statements from the CEO/COO and Director/VP for Safety or Quality Assurance).
- Accident/incident investigation reports (how are undesirable events investigated and what are the outcomes?).

Safety Strategies may be discovered in organization-wide programmes, like the Safety Management System, or through the extant degree of alignment between safety values and the various business processes, policies, and procedures. The degree to which top leaders reinforce safety values is an important data point. Also, mechanisms in place to effect systemic improvements, such as non-punitive error reporting systems, pre-/post-procedure briefing and debriefing, a lessons learned database that feeds technical and operational training, employee socialization programmes, etc. provide clues about the relative value placed on safety. Typically, we find organizational structures, policies, or processes changing in response to specific unsafe events. For example, a procedure is changed because the original procedure was erroneous or prone to errors, but rarely do we find a comprehensive process to evaluate all similar procedures or to improve the quality assurance programme such that all procedures are evaluated periodically. In some cases, we have found that procedures were not tested prior to initial implementation and embedded errors continued until they manifested in undesirable outcomes.

An assessment of strategies can also provide valuable insight into the state of organizational learning – how does the organization learn and how is that learning documented or retained? When accidents/incidents from similar causal factors continue to occur, it is evidence of failure to learn. If the changes in response to undesirable events are episodic or narrowly limited to the specific procedures, the learning may be intermittent. When an effective quality assurance system is able to demonstrate continuous assessment of the effectiveness of safety strategies, the organization may be in a state of continuous learning. The highest state of learning, transformative learning, exists when the organization is proactive in its learning, engages in deep reflective practice, and also influences industry-wide changes. Again, the leadership's personal commitment to safety and their ability to effect strategic changes in the organization are key factors in determining how far the organizational culture will change.

Safety climate is most commonly assessed with safety attitudes and opinions surveys. Since these surveys are snapshots of employee attitudes and opinions, results from just one survey are not very useful; they need to be conducted at regular intervals for several years. As the survey results are compared across multiple measurement cycles, a meaningful safety culture profile can be constructed. Such surveys are particularly effective in indicating effects of specific safety culture interventions. When safety performance is also tracked along the same time-frame, a causal/contributory link may be made between safety climate and safety performance. In this sense, safety climate surveys could be used to test the effectiveness of particular safety strategies on safety attitudes and behaviours.

In both aviation and healthcare, safety performance is being tracked from the perspective of broad national or international goals that are customized to local application. The tie between organizational safety goals and the national/global goals provides a sense of anchoring or stability and reduces the tendency of field personnel to think of the organizational safety goals as temporary management fads that they can ignore. Some organizations have taken bold steps at declaring ambitious goals and followed with building structures, processes, performance metrics, and resources allocation formulae to achieve the stated goals. Examples presented in this book indicate that safety performance can be managed at national as well as local level, but the goals need to be measurable, realistic, and supported through resources, and there has to be a commitment to change organizational structures, reward systems, and attitudes toward errors.

Safety Culture Transformation

Safety Culture Transformation is about changing an organization from the inside out. The transformation process challenges the validity of organizational values and seeks to change at least the enacted values, when necessary. Such a deep change is not for every organization and it is not easy to accomplish. It requires a solid commitment from the Board, the CEO/COO, and all the way down the management chain. In aviation, this commitment includes labour unions and the regulator. In healthcare, there has to be a commitment from physicians, who may not be employees of the hospital, just as there needs to be a commitment from the hospital leadership and staff. However, with top-level commitment and sustained efforts, it is possible to transform the safety culture of any organization. Often, such a transformation impacts not only the safety performance, but also the business performance and employee and customer satisfaction.

The first, and a recurring, step in safety culture transformation is safety climate assessment. A baseline measurement of safety climate prior to the launch of any transformation efforts will help compare the shift, if any, in safety attitudes and opinions as the transformation initiatives are implemented. The actual implementation efforts may start with a small, contained project like the implementation of a non-punitive error-reporting system in just one section of the organization. Another approach would be to announce the overall goals, but let the various organizational units develop their own initiatives to achieve the broader organizational goals. The specific path chosen would depend on the nature of the business and the organizational readiness for change – some organizations may be more receptive to a centralized model while others may be more receptive to a decentralized model.

Conclusion 195

Safety performance and financial metrics need to be tied together. Errors in both aviation and healthcare are astronomically expensive, yet the processes used to track the financial impact of such errors are woefully inadequate. Once safety performance and financial metrics are tied together, it would be easier to determine not only the cost of a particular type of error, but also the cost avoided over a particular performance period. Next, financial mechanisms need to be built so that a portion of the cost avoided is routed back to the maintenance and development of the safety initiatives. Once the managers see that cost avoidance has tangible benefits to their operating budget, they will be more enthusiastic about implementing and protecting safety programmes. Safety programmes represent a financial *investment* in the organization. As such, they need to generate a reasonable return on investment; the more effective the safety programmes, the greater will be the need for investment.

Next, we present two comprehensive examples illustrating vertical and horizontal analyses of safety culture in accordance with the Safety Culture Pyramid and transformation across the various states of safety culture.

Example 8.1: ATO Technical Operations

The Federal Aviation Administration (FAA) provides air navigation services through the Air Traffic Organization (ATO). The ATO was created in 2004 to respond to increasing air traffic capacity demands and to enhance the safety and security of the National Airspace System (NAS). Approximately 25 per cent of the ATO's 35,000 employees work in the area of Technical Operations (Tech Ops) and are responsible for installing and maintaining equipment used by air traffic controllers, who serve commercial, private, and military aviation. This case was part of an FAA funded research study (Patankar and Sabin, 2008) aimed at understanding and transforming safety culture in Tech Ops. It started with a fundamental question asked by senior management: 'what is safety culture like in Tech Ops?'

Overview of Safety Culture Assessment Methods

The ATO Tech Ops Project was designed as a research study and used multiple assessment methods to understand safety culture. The project began with a comprehensive review of the existing research literature on safety culture. The second step involved using a variety of qualitative methods such as interviews, focus groups, field observations, and artifact collection, to create a preliminary understanding of existing policies, procedures, practices and dynamics surrounding safety. Third, based on the qualitative findings, a questionnaire was designed and distributed to measure the state of safety climate and provide a baseline for pre-post intervention comparisons. Fourth, an intervention (a non-punitive error reporting system) was designed and implemented to improve safety culture. Errors identified through the reporting system were subjected to a comprehensive and systematic case analysis to determine the contributing causes, corrective actions required, lessons learned and best practices for dissemination to improve safety. Fifth, to evaluate the impact of the intervention on safety culture, a quasi-experimental design was used to assess differences in safety culture between control sites compared to demonstration sites that received the error reporting system intervention. Throughout the three-year project, frequent progress reports, interim analyses, and in-depth dialogues with Tech Ops management and staff helped develop a shared understanding of the findings and shaped subsequent stages of the project.

Results

Review of the research literature The safety culture literature review was designed to answer questions of particular interest to the ATO: are there examples of organizations that have strong safety cultures from their beginning? Are there examples of organizations that have successfully transformed their safety cultures? If so, how did these organizations do it? What attributes changed? How was the change measured? What resources were needed? What obstacles had to be overcome? What were the benefits of the transformed safety culture?

To address these questions, the literature review included studies conducted across a variety of high-consequence industries such as nuclear power, chemical manufacturing, railway systems, petroleum operations, construction companies, healthcare, the US Navy, and US aviation (Patankar, Bigda-Peyton, Sabin, Brown, and Kelly, 2005). The review demonstrated that there were very few organizations that had strong safety cultures from the start. One example of a strong safety culture by design is the Diablo Canyon Nuclear Power Plant. Most industries, however, have learned to become safer over time by explicit changes that promote safer operations. Cross industry comparisons have resulted in the development of robust theories related to safety such as the Normal Accident Theory (Perrow, 1984), High Reliability Theory (La Porte and Consolini, 1991; Roberts, 1993), Ultrasafe Theory (Amalberti, Auroy, Berwick, and Barach, 2005), and Blame, Reporting and Just Cultures (Reason, 1997; Marx 1997, 2001). Concepts from these theorists helped frame the Tech Ops study and provide a foundation for its approach.

Likewise, programmes, methods, metrics, techniques, lessons learned and success stories especially from the US Aviation Industry offered potential intervention strategies for the ATO to consider. Foremost among these were the Aviation Safety Action Program (FAA, 2002), Flight Operations Quality Assurance (FSF, 1998), and Crew Resource Management (Helmreich and Merritt, 1998). The concept of organizational learning provided an integrative process for understanding how teams and institutions can profit from experience and change culture (e.g. Argyris and Schön, 1974). Lastly, a set of team, organizational, and outcome factors were identified that have contributed to safety culture previously and were hypothesized to also impact Tech Ops (Patankar, et al., 2005).

Qualitative results Structured interviews and focus groups using a standard protocol were conducted at FAA/ATO headquarters and six facilities representing the three geographic service areas. Over 75 individuals participated in this data collection. With the consent of participants, most sessions were digitally tape recorded and later transcribed to ensure accuracy of the data. Transcripts were content analyzed to identify major themes and supporting examples.

Key findings from frontline personnel included the following: safety in Tech Ops was heavily dependent on the frontline employees' strong sense of professionalism, commitment to operational reliability, and a feeling of ownership and personal responsibility for the equipment that they maintained. Employees demonstrated a great sense of pride in their accomplishments and satisfaction with their work. However, frontline employees noted conflict between maintaining safety for the flying public and increasing pressure to expedite the flow of air traffic. Some who were familiar with the *Columbia Accident Investigation Board Report*, questioned whether the FAA's safety culture is heading toward the NASA Culture of 'faster, better, cheaper', thus implying that an erosion of the institution's core values and identity was already underway. Field personnel often maintained that senior leadership's words speak of safety but their actions are all about efficiency: they conclude 'Headquarters does not practice what it preaches about safety'. This led many employees to question where Tech Ops was heading in the future.

The Tech Ops field-level work environment was marked with discontent regarding decreasing staffing levels, diminished standards for many new hires, declining experience-level of the workforce due to retirements, and difficulties in securing the appropriate levels of training. Field managers often reported that they were held highly accountable by headquarters to use limited fiscal and human resources to get the job done. Many personnel also noted serious communication disconnects between the local, regional and headquarters levels. Headquarters was often seen as increasingly out of touch with local issues and unresponsive to employees and managers in the field. It was commonly stated that no one higher up the chain of command wanted to hear about a problem in the field.

Many field employees reported participating in planning for improvements, but their recommendations were seldom acknowledged or implemented. As a result, personnel were reluctant to participate in new initiatives and often regarded new programmes from headquarters with cynicism as just another 'flavor of the month' that would soon disappear. Many argued that due to the FAA's bureaucracy, change would not happen until an accident occurred that could be traced back to a compromised safety standard. Tech Ops field employees pointed out that the FAA is known as the 'Tombstone Agency' and commented, 'Once people are killed, things will change and more staff will be hired'.

While employees often stated that they trust their immediate supervisor there was much less trust in higher levels of management or senior leadership. Field employees often claimed that they were not comfortable reporting safety issues to management because of fear of reprisal; instead, they would either correct the errors themselves or hope that these errors were not discovered. Employees also gave examples of managers falsifying data in order to protect certain individuals or facilities. From the senior management's perspective, multiple reporting systems existed, but they were not being used effectively. From the employees' perspective, reporting systems such as the 'Lessons Learned' database were one-way repositories of information rather than a learning tool. Employees who made a mistake often viewed writing the 'lesson learned' report as a punishment. If a system-wide version of the report was to be disseminated it was often 'sanitized' to remove critical details for political reasons.

Tech Ops field personnel reported that they often needed to do workarounds to compensate for lack of component compatibility or out-of-date procedures. They maintained that they were seldom consulted on equipment procurement, but they frequently needed to install and maintain new equipment before it was adequately tested and ready for deployment. It was very clear that Tech Ops field personnel were extremely dedicated to NAS safety. They have a 'can do' attitude, but their ability to maintain the safety of the system is expected to reach a critical point in the near future with an increasing number of 'unreliable' technical systems being fielded that are to be maintained by a shrinking workforce; many viewed their work becoming a 'mission impossible'.

Tech Ops personnel operate under a complex set of performance pressures related to limited staff, tight budgets, and strict accountability for equipment maintenance to protect NAS safety. To accommodate these demands some managers had secretly changed ('pencil whipped') outage reports to improve facility performance metrics (e.g. system reliability and availability measures). One Tech Ops field employee succinctly captured the web of systemic relationships between this secretive culture and NAS safety:

> An unscheduled outage may be changed to a scheduled outage to make the facility look better; however, the result is that equipment appears not to need preventative maintenance because it did not fail. If equipment is not prone to periodic failure, then less manpower is needed to maintain it and new hires are not made. This creates a culture based on lies.

Ultimately in this scenario, systems failed, which exposed the NAS to unnecessary risk because there were not enough Tech Ops specialists to provide the preventative maintenance that was actually required.

Pre-intervention questionnaire results Based on the qualitative findings and guided by theory from the literature review, the *ATO Technical Operations Safety Climate Questionnaire* was developed. This questionnaire consisted of 58 items that were rated on five-point Likert scales (strongly disagree to strongly agree). The questions were organized into four main factors with 16 subscales and included the following: Organizational Factor (Institutional Identity, Information Flow, Relationships, Leadership, Evaluation, Error Management); Team Factor (Supervisor Trust, Goal Sharing, Professionalism, Adaptability, Support Systems,

Quality Management); Technical Factor (Technical Standards); and Outcome Factor (Employee Satisfaction, Customer Satisfaction, Safety Performance). The purpose of the questionnaire was to quantify safety attitudes and opinions. Findings provided a snapshot of the current state of employee perceptions of safety climate in Tech Ops. The questionnaire data allowed for comparisons between sites and service areas and provided a baseline measure of safety climate that subsequent re-administrations of the questionnaire could be compared with in order to determine changes over time.

The questionnaire was distributed to two Tech Ops locations in each of the three geographic service areas (western, central, and eastern). A total of 180 Tech Ops personnel completed the survey, which represented an overall response rate of about 65 per cent. At the factor-level analysis, the Outcome Factor was rated most favourably (mean rating of 3.79/5.00). The Team Factor ranked in second place (3.65/5.00), the Organizational Factor ranked third (3.15/5.0), and the Technical Factor ranked in last place (2.52/5.00).

Figure 8.1 presents the 16 subscales ranked from most to least favourable in terms of their mean ratings. The subscales provide additional insight into the employees' perceptions of safety climate and corroborate qualitative findings from the interviews and focus groups. Professionalism, Goal Sharing, and Employee Satisfaction subscales ranked most favourably and confirmed clear themes from discussions with Tech Ops personnel regarding their work ethic, professional pride, and commitment to the goal of maintaining NAS safety. Tech Ops'

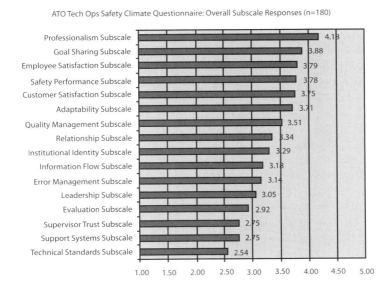

Figure 8.1 **Tech Ops response to safety climate survey**

mediocre ratings of the Organizational subscales reflected concerns expressed in interviews and focus groups about the quality of leadership, information flow, error management, and performance appraisals as well as ATO's inconsistency between words and actions in relation to safety. The frequency of comments from focus groups regarding limited resources, and staffing and training inadequacies, as well as concerns over the number of required workarounds coincided with the lowest subscale ratings for Support Systems and Technical Standards. Employees also reported a lack of safety-related discussions during employee performance evaluations that were regarded by many 'as a joke' due to their superficial nature. System safety was largely due to the professionalism of individual specialists working in an environment where there was a general lack of technical standardization. In conclusion, the pre-intervention questionnaire results demonstrated that there were multiple areas and opportunities for the improvement of safety culture in Tech Ops.

The safety culture continuum The researchers were asked to identify the overall existing state of safety culture in Tech Ops. Dominant states of safety culture previously identified in the literature were used to analyze and organize observations. The researchers found variations between different sites and between the safety perceptions of personnel working in the field compared to those at headquarters. Throughout the research process examples were found of Blame Culture, Reporting Culture, and a desire for Just Culture. These dominant cultural types can be arranged on an accountability continuum that represents different levels of organizational maturity in addressing safety issues. Of particular note was the discovery at some locations of a new cultural type, which the researchers labeled Secretive Culture. The norm in secretive culture was concealment of mistakes that did not come to the attention of superiors and that did not result in recognizable harm. Additionally, evidence was found of 'pencil whipping' some reports so that errors were hidden and performance metrics of system reliability and availability remained at high levels. The Safety Culture Continuum is presented in Figure 8.2.

Qualitative and quantitative data were combined to position the markers on the various dimensions for the field-level workforce compared to senior-level management. The greatest discrepancy in the perceptions of safety culture occurred for the dominant state of safety culture along the accountability scale. Senior-level management believed that adequate processes and procedures were in place to support a reporting culture. However, the field-level workforce believed that the existing culture was one primarily based on blame with some secretive subcultures (left of blame on the accountability scale). There was agreement between both groups that Tech Ops' approach toward learning was intermittent.

A non-punitive error reporting system was selected as an intervention to improve safety culture in Tech Ops. This decision was based on evidence that this approach had been effective in bringing about positive change in the past. Additionally, commercial aviation uses a robust error reporting system, which

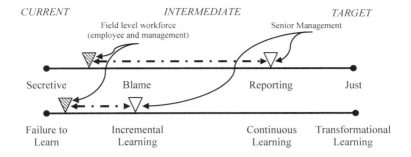

Figure 8.2   States of safety culture in Tech Ops

seemed transferable and relevant to Tech Ops. A quasi-experiment was designed to test the effectiveness of this intervention.

Design and implementation of the non-punitive error reporting system   The basic intent of establishing the error reporting system was to stop affixing blame and start solving problems by moving away from a secretive/blame culture and toward a reporting culture. The basic rationale for paying attention to errors instead of hiding or ignoring them stems from the work of H.W. Heinrich (1931) who recognized that for every serious accident numerous precursor events including unsafe acts and near misses had occurred. By reporting and understanding the deeper systemic causes associated with these less serious events, knowledge may be acquired and applied to prevent large scale events that cause serious damage, injury or loss of life.

A non-punitive error reporting system was designed for ATO's Technical Operations employees and managers. This system, named the Tech Ops Aviation Safety Action Program (ASAP), was based on the fundamental principles and the best practices learned from the airlines' use of ASAP (FAA, 2002, AC 120-66B). Tech Ops ASAP was implemented at three demonstration sites with one located in each of the three geographic service areas. A control site that had participated in pre-intervention focus groups and surveys was also available for comparison in each service area. Figure 8.3 presents a model of the programme.

Key steps in the ASAP reporting system included the following:

- ASAP reports were submitted from the field directly to Saint Louis University. Reports that pertained to a specific error committed by the reporter or the recognition of a general hazardous condition that could impact NAS safety were accepted for further consideration; others were rejected.
- Reports were received and evaluated by ASAP researchers at Saint Louis University. If the report met the acceptance criteria it was de-identified and forwarded to the Event Review Committee (ERC), which consisted of Tech Ops management and employee representatives. Reports were not accepted

if the reporter acknowledged the commission of a crime, an error while under the influence of drugs or alcohol, intentional disregard for safety, or sabotage. If the report pertained to an urgent or continuing serious threat to the NAS, the issue was immediately referred to Tech Ops Management.
- Reports transmitted to the ERC were discussed and a variety of contributing factors were analyzed: environmental factors, documentation, training, hardware, and organizational factors. Ultimately, the ERC decided, by consensus, whether the error was an honest mistake or reckless behaviour. The ERC also established consensus on the recommended solution.
- If the error was judged to be an honest mistake, the report was accepted and a change action recommendation was forwarded to the appropriate programme office.
- If the error was judged to be an example of recklessness and not accepted by the ERC, a communication was sent back to the reporter, and if safety recommendations were made, they were communicated to the appropriate management representative.
- ASAP researchers at Saint Louis University sent feedback letters to the reporter with information regarding the ERC's final decision, its recommendations, and in some cases, the status of the recommendations.

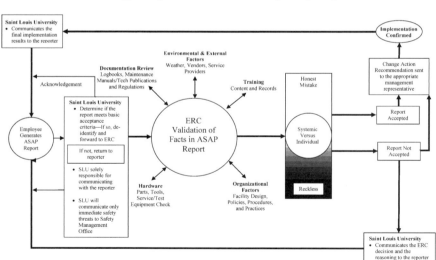

Figure 8.3   The aviation safety action programme in Tech Ops

At the heart of this process were confidential protection for reporters, an Event Review Committee (ERC) that thoroughly investigated the de-identified cases, protection of reporters from repercussions if the case was judged to be an honest mistake, recommendations for improvement to correct systemic causes of error, and dissemination of findings to prevent similar errors in the future. The ERC committee members served as both evaluators of the ASAP reports and champions of the programme. They led a series of informational presentations, distributed ASAP posters, and wrote newsletter articles to explain the process and market the concept of ASAP to the respective constituencies.

Removing barriers to using the reporting system   In early ASAP briefings that introduced the programme, many Tech Ops employees gave a surprisingly cynical response. They viewed ASAP as a 'get out of jail free card' with potential benefit limited to less proficient employees. The non-punitive aspect of ASAP was viewed with suspicion and did not align with the specialists' historical experience with the dominant command and control, blame-oriented culture. From further dialogue it became clear that engaging the employees' strong sense of professionalism could increase ASAP's appeal. Employees who became champions of the process discussed the value of voluntarily reporting their errors and viewed this as a professional responsibility that was consistent with their strong work ethic and saw it as an essential part of their professional identity to ensure NAS safety. Receptiveness to the ASAP programme improved when it was introduced as a means to encourage a new professional norm. Viewed from this frame of reference, reporting of unintentional errors seemed more reasonable to the Tech Ops employees.

However, during the first several months of the programme only one error report was filed. To identify barriers to reporting, the research team engaged employees from the demonstration sites in conversations about ASAP. Primary barriers to reporting included the following: insufficient knowledge about the programme and its purpose; limited motivation of Tech Ops specialists to report mistakes due to the desire to protect their jobs; limited written commitment from management; lack of engagement from the labour unions; and fear of confidentiality being breached due to the relatively small number of potential reporters. To address these barriers and increase participation in the programme, additional briefings were held which included information on management's support, and the scope of the project was expanded to include additional sites. These efforts increased reporting and a total of 11 ASAP reports were filed over the 18-month duration of the project. These reports and the corrective actions that were implemented impacted the ATO at the local, district and national levels.

Post-intervention questionnaire results   A post-intervention survey of safety climate was administered at the three control sites and the three demonstration sites (including the expanded sites). The goals of this survey were to determine if safety climate had improved as reflected by employee attitudes and opinions, to identify if statistically significant differences existed after implementation of the ASAP programme, and to document any barriers and/or enablers related to the reporting of errors.

To better assess ASAP, 28 additional Likert-type questions and four open-ended comment questions were added to the pre-intervention questionnaire. A total of 164 questionnaires were completed: 116 responses were received from the ASAP demonstration sites and 48 were returned from the control sites. Since the number of responses from the control sites was too low for meaningful analysis, pre-post intervention comparisons were based only on the intervention sites.

Pre-post comparisons of the safety culture questionnaire data were made for the ASAP intervention sites. Fourteen of the original sixteen subscales showed improvement: the magnitude of positive changes for Institutional Identity and Information Flow scores were statistically significant $(p < .05)$. Two of the subscales showed decline: the decrease for Technical Standards score was statistically significant $(p < .05)$.

ASAP safety culture briefings were attended by over 475 Tech Ops personnel during the demonstration phase of the project. It appears that this communication improved perceptions of information flow up and down the chain of command and affirmed ATO's core mission of safety. At the same time, these briefings increased awareness for Tech Ops personnel of the hazards posed by the lack of sufficient equipment compatibility, the abundance of workarounds, and threats posed by the secretive/blame culture leading to a reduction in ratings for Technical Standards. The consistency of improvement among 14 of the 16 safety culture subscales during the 18-month period was very encouraging. This may indicate the unfreezing of the existing secretive/blame culture, heightened awareness of the causes of unintentional error, and movement toward a new professional norm supporting a reporting culture.

The statistical technique of multiple regression was used to identify questionnaire items that predicted agreement with the statement: 'Tech Ops ASAP should be implemented at the national level'. The strongest predictors included the perception that ASAP is a genuine attempt to improve safety, handling reports in a prompt manner, employee-manager trust, and assuring confidentiality of the reporter. These issues represent important points of leverage to enable the effectiveness of ASAP and movement towards a reporting culture. If these issues are not sufficiently addressed they serve as programme barriers.

Lessons learned from the Tech Ops project One of the key transformational lessons learned over the course of the Tech Ops project was how changes in assumptions, attitudes and behaviours impact safety culture. Very few people could have imagined the possibility of a programme like ASAP ever coming to fruition, especially *within* the FAA's bureaucracy. Initially the idea of voluntarily self-reporting errors was unthinkable for many. Over the course of three years, numerous briefings and conversations were held with members of ATO to reflect on safety culture, organizational mission, leadership strategies, core values, and deeply held beliefs. Concurrent with these ATO discussions was the widespread, international recognition of safety culture and its significance in improving global aviation safety. These forces combined so that layer after layer of FAA management started to believe in the possibility of cultural change and the benefits of learning from a non-punitive error reporting system. This increasing belief in the benefits of a better safety culture was genuine and was passed on from the VP of Tech Ops to all the reporting layers of management, from headquarters to the field, and to individual frontline employees. Consequently, attention, traction, and support for the programme cascaded across every level of the organization.

One notable deficiency in the research programme was that the labour union did not take a formal or active role. The research team discovered a level of readiness in members of the Professional Airway Systems Specialists (PASS) union and in ATO managers to work together and build a non-punitive error reporting system. Nonetheless, communication across the labour-management lines was very challenging due to a variety of issues. These issues need to be addressed prior to moving forward with programme implementation at the national level.

Conclusion

Safety culture can be improved by intentional interventions. The Tech Ops story demonstrated that a non-punitive error-reporting system can move a secretive/blame oriented culture toward the preferred states of reporting and just culture. The reporting system was effective in identifying actionable solutions to systemic issues that mitigated risk and improved safety of the NAS. Establishing the reporting system as a professional responsibility was an important first step in improving transparency and increasing safety. This research study benefited from the use of multiple methods to assess safety culture. Questionnaires measured underlying dimensions of attitudes and opinions related to safety climate and provided a baseline for longitudinal and intergroup comparisons. Interviews, focus groups, and dialogue enabled a rich description of the culture's values, strategies, norms, underlying values and assumptions. Case analysis of error and hazard reports allowed a systematic investigation of events, identification of contributing causal factors, the generation of solutions, and the dissemination of best practices. The quasi-experimental method examined the impact of the error reporting system on changing safety culture by comparing data from the control and demonstration sites. These diverse methods provided complementary information that yielded a comprehensive understanding of safety culture.

The benefits of the current approach could be applied to a variety of high-consequence industries that wish to improve safety culture. However, cultural change requires a significant investment of time, effort, and resources. This research also demonstrated the importance of being mindful of powerful influences exerted by the social and organizational context (e.g. leadership commitment, goal sharing, trust, and communication) that impact change efforts.

Moving forward, the entire Air Traffic Organization, of which Tech Ops is a part, is committed to building a stronger safety culture. Figures 8.4–8.6 present artefacts that illustrate institutional commitment to this effort. The Air Traffic Controller Aviation Safety Action Program's Memorandum of Understanding (ATSAP MOU) was signed on March 27, 2008, and it launched the Air Traffic Controller ASAP programme. This is a non-punitive error reporting programme, much like the airline ASAP and also similar to the Tech Ops ASAP that was discussed earlier (the Tech Ops ASAP programme was limited to the research project and was not considered to be operational). The ATSAP MOU was the essential first step in moving toward a reporting culture; it illustrates the following:

- The labour union (National Air Traffic Controllers Association) and the FAA believe in the value of error reporting.
- The MOU offers an implied commitment to address systemic weaknesses discovered through the ATSAP system.
- Three key groups of the organization – labour, management, and regulator – are committed to building a non-punitive error reporting system; they are collectively striving toward a reporting culture.

Figure 8.4 illustrates a poster used to raise the awareness regarding incidents, hazards, and unsafe acts ultimately leading to an accident. The image of a child looking through the window is intended to maintain focus on the safety of the flying public. Figures 8.5 and 8.6 illustrate the commitment from the highest officer in the ATO, the Chief Operating Officer, and his direct report, the VP for Safety. Together, they personify the institutional commitment.

Figure 8.7 demonstrates a public commitment toward improving safety culture. Specific goals and timelines are listed, and this document is available on the employee website at <https://employees.faa.gov/org/linebusiness/ato/safety/>.

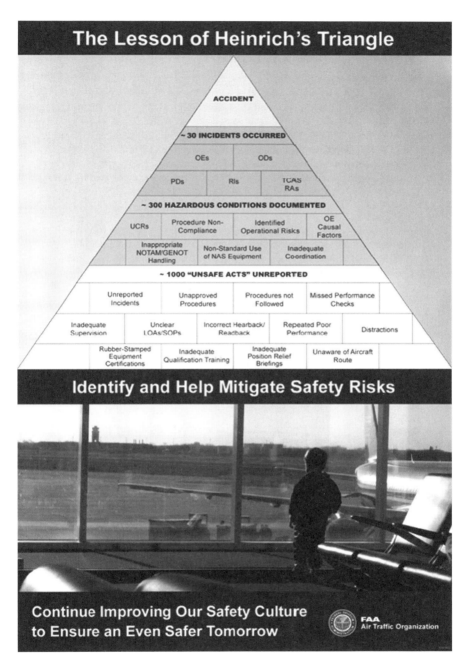

Figure 8.4　The ATO Heinrich triangle

Foreword

The safety of National Airspace is our most important priority. Our objective is to continually improve our safety performance while we transition to the Next Generation Air Transportation System (NextGen). To do this, we will need a strong safety management system.

History demonstrates that we must be poised to handle future demand that will surely return as the nation's economy improves. In fact, the aviation sector will be an important factor in the nation's economic recovery. The FAA estimates that in 2006, civil aviation accounted for 11 million jobs and represented 5.6 percent of the gross domestic product. At $61 billion, aerospace products and parts contributed more to positive balance of the trade than any other sector—$32 billion more than the next highest contributor.

With so much riding on our service, we are constantly working to increase the safety of an already exceedingly safe system. To do this, we are making a transition from the traditional forensic investigations of accidents and incidents to a more prognostic approach to improving safety. We are committed to building a positive safety culture, with open exchange of pertinent safety information. Transparency has never been more important.

This Safety Blueprint outlines our commitment to safety during this critical transition.

Sincerely,

Hank Krakowski, Chief Operating Officer
Air Traffic Organization

Figure 8.5   A statement of commitment from the chief operating officer 'the safety blueprint', April 2009. Available from <https://employees.faa.gov/org/linebusiness/ato/safety/>

5.0 Office Of Safety

Established in 2004, the Office of Safety is delegated the primary responsibility for safety assurance within the ATO. The Office of Safety is responsible for ensuring that all ATO service units integrate safety responsibilities into their provision of service.

The Office of Safety works with operational service units to lead ATO efforts to manage risks, assure quality standards, instill an open culture of disclosure, educate employees and promote continuous improvement.

Responsible for identifying and mitigating aircraft collision risks during the delivery of air traffic separation services, the Office of Safety is the focal point for:

1. Applying the agency's SMS principles.
2. Auditing safety, quality assurance and quality control in the ATO and reporting findings to improve safety performance.
3. Integrating the functions and information of risk reduction, investigations, evaluations, independent operational testing and evaluation, safety risk management, runway safety and operational services, in order to identify collision risks and influence their resolution.
4. Providing information on assessments of operational and safety performance within the NAS.
5. Working with the Associate Administrator for Aviation Safety and other external entities undertaking special projects in support of increasing the safety of the NAS.

6.0 Conclusion

The Office of Safety will strive to work closely with ATO service units to identify additional safety initiatives. This Blueprint for Safety will be updated annually to ensure our efforts reflect the agency's safety priorities.

> "Moving forward, we must be the catalyst for creating a **fundamental safety culture** throughout the ATO. We want each individual at every level of the ATO to hold safety as an **enduring value and priority.**"
>
> — Bob Tarter, Vice President Office of Safety

Figure 8.6   A statement of commitment from the VP for safety

Safety Culture Initiatives: 2010 and Beyond

Attribute	Safety Strategy	ATSAP	QA-QC	Credentials	New Metrics	SMS
Informed Culture	Informs our employees and stakeholders of safety objectives and priorities	Expands knowledge of safety events and provides feedback loop that reinforces reporting	Clarifies safety responsibilities between facilities, Service Centers and Office of Safety	Increases awareness of certification process and recurrent training requirements	Introduces metrics based on risk; enables enhanced and standardized causal factor analyses	Managers have current safety knowledge of the system
Reporting Culture	Recognizes the value of employee input	Event Review Committee (ERC) corrective action requests reinforce value of reporting	Eliminates duplication of effort; reduces operational error emphasis; makes better use of resources; develops mitigations locally	Enables non-punitive reporting with cooperation of our regulator	Enables electronic reporting of events and encourages use of analytical tools and human reporting	Data integration enables complex analyses across multiple databases (e.g., ASIAS)
Just Culture	Demonstrates commitment to non-punitive safety reporting	Disconnects report from performance management Program grows with trust	Division of work (QA-QC) builds organizational trust	Rewards reporting; clear about unacceptable behavior	Holistic approach to risk raises our credibility and removes reporting stigma	Creates atmosphere of trust; rewards reporting; clear about unacceptable behavior
Learning Culture	Supports internal awareness of COO's Safety Vision	Systematic findings of mitigations/best practices are communicated and implemented widely	Changes QA focus to system evaluation and analyses Develops recommendations based on best practices; informs integrated training development	Demonstrates organizational commitment to recognizing individual achievements and targets recurrent training	Applying resources to highest risk events will yield links between causal factors and risk	Modeling and simulation laboratory increases our understanding of risk and enables continuous improvement; enhances quality of integrated training
Flexible Culture	Integrates ATO's safety objectives to facilitate major reforms	Corrective actions are implemented quickly using mitigations/best practices in use elsewhere	Enables new mission without increase in resources	Enables better emergency response based on electronic knowledge of pervious certifications	Focused analysis leads to faster changes in a high tempo environment	Accumulated safety case knowledge allows faster and better changes in the system
International Harmonization	Identifies common values and objectives	Information sharing adds to ATOs body of knowledge	Standardizes the ATO with other industry models; e.g. ISO9000, airlines, etc.	Aligns FAA and ICAO requirements	Aligns FAA with major Air Navigation Service Providers (ANSP) metrics	Aligns FAA and ICAO requirements; cross-border cooperation leads to information sharing of safety cases and increased efficiency of resources

Figure 8.7   Safety culture initiatives commitment

Example 8.2: North York General Hospital

Background and Context

North York General Hospital (NYGH) is a major hospital north of Toronto, Canada. In 2003, the hospital found itself at the epicenter of Severe Acute Respiratory Syndrome (SARS). This was a shattering and traumatic event for the community, including quarantines, closing of sections of the hospital, prolonged emergency conditions, and loss of life (including a nurse). In the wake of the epidemic, the hospital was shaken, demoralized, and disoriented.

Viewed from the perspective of the Safety Culture Continuum, North York General has undergone a transformation of culture, leadership, and performance since 2003 which has moved the hospital from a state of Blame Culture to one of Reporting Culture, with early signs of a shift to Just Culture as the normal operating environment. This movement has been accompanied by parallel evidence of performance improvement, as shown by staff and patient safety surveys, improvements in patient flow and wait times, and a significant decline in adverse events, such as ventilator-acquired pneumonia and central line infections. In addition, there has been considerable qualitative evidence of the shift, including increased alignment between the Board, administration, and staff; the growth of a proactive quality, risk, and safety group; and growing collaboration across service units, for instance with respect to patient flow, wait times, and discharge planning. At the same time, the North York journey is far from over. As of this writing, the hospital is pursuing a new era of patient safety improvement, with a goal of improving organizational resilience, internalizing the new culture in everyday practice and operations, and achieving ultrasafe levels of safety performance (a standard previously attained only in high-consequence industries outside of healthcare, such as chemical, nuclear, and aviation).

Safety Performance at NYGH

Applying the Safety Culture Pyramid to the experience of North York General, we will begin with 'Behaviour,' or safety performance. Starting in 2003–2004, this kind of performance was measured in three different ways:

- Indirectly, by using the Patient Safety theme from the Balanced Scorecard.
- Directly, by examining adverse events and near-misses.
- Prospectively, by a collaboration between the Board and administration to come up with metrics which the Board Quality Committee could use to evaluate performance in quality and safety.

In addition, starting in 2008, the hospital began using a 'Safety Drivers' model that identifies key 'drivers' (or success factors) for improving patient safety and organizes assessment, initiatives, and evaluation around these factors.

Conclusion

Efforts to improve safety at North York General were not prompted by an adverse event or by a perceived performance problem. Rather, they were prompted by the following factors:

- A forward-looking CEO, who was also personally committed to patient safety, and a progressive Senior VP of Quality who was involved with the Institute for Healthcare Improvement (IHI) and was committed to the 100,000 Lives Campaign/Safer Healthcare Now campaign and associated practices such as the review of critical incidents involving patent safety and quality.
- A culture of innovation, initiative, and teamwork that was promoted, nourished, and sustained during this era.
- A Board that respected the CEO for many things, including her vision, innovation, and ability to engage front-line staff and the wider community, but was also nervous about what they perceived as a lack of 'hard' metrics with which to evaluate the evidently positive, but also 'soft', changes taking place at the hospital.
- A climate in the wider healthcare industry, which suggested that patient safety was a critical issue and needed to be improved, that the managing boards should be held accountable for patient safety, and that critical incidents, such as adverse events, should be discussed openly and reviewed for possible learning rather than avoided or made occasions for affixing blame.

Prior to focusing on quality, safety, and risk, the Board at NYGH had worked for some time on a Balanced Governance Scorecard and had become accustomed to identifying targets, indicators, goals, and metrics. When they began what later became known as the Board Quality Journey, they were working from this perspective.

One other important environmental factor was the priorities of the Ministry of Health. During this period, a key priority was patient flow and wait times, focusing on Emergency Departments (ED) and General Internal Medicine (GIM). These were not seen as patient safety initiatives, but at NYGH the administration and staff went through a process/story-mapping exercise that helped them to identify the unintended consequence for patient safety of periodic surges in the patient population and the bottlenecks that resulted. This led to a significant effort to reduce cross-unit variation in practices and processes, greater emphasis on discharge planning, and heightened efforts to coordinate with local and community care facilities, most recently through the Community Care Access Centre in the North York region.

The most important driver of improvements in quality, safety, and risk management at NYGH during this era was the internal impulse toward innovation at the hospital. This was manifested in many ways. For instance, it is noteworthy that NYGH was working on the kinds of improvements advocated by IHI,

Canadian Patient Safety Institute (CPSI), the Board, and the Ministry of Health well before these priorities became visible, popular, or attached to rewards and punishment. Throughout this period, NYGH administration and staff were early adopters of new improvement practices. Their motivation was far more intrinsic than extrinsic. As a consequence, the hospital became recognized as a great place to work, a hospital of choice for patients, and a leading community hospital as recognized by the wider community and the Ministry of Health.

Referring back to the Safety Culture Pyramid, safety performance during this period at NYGH was viewed as closely linked to performance in Quality and Risk. It was evaluated through the following lenses:

- The Patient Safety/ Patient Experience Themes in the Balanced Governance Scorecard.
- Specific goals and monitoring established by the Board Quality Committee, working in conjunction with the administration and staff (note: this was based on the Board Quality Committee asking Susan Kwolek, Sr. VP of Quality, for her 'top five' priorities within the Scorecard and her portfolio, then coming up with indicators to measure progress on these.).
- Metrics defined by the 100,000 Lives Campaign led by IHI (in Canada, this became the Safer Healthcare Now initiative, sponsored by the Canadian Patient Safety Institute).
- ED-GIM performance (wait times and patient flow in the Emergency Department and General Internal Medicine), as defined by the Ministry of Health.

In addition, the definition of 'performance' in patient safety and quality was undergoing continuous reflection and evolution in the light of emerging areas of interest to senior leaders and staff, the use of intuition in clinical decision-making, and the development of emotional intelligence as a characteristic of well-performing managers and leaders.

Safety Climate at NYGH

The next layer of the Safety Culture Pyramid is Safety Climate (attitudes and opinions). As noted previously, North York General had set a goal of becoming a 'learning organization' in 2003. The hospital worked with a model of transformation called the Quantum Journey, designed and led by Ted Ball of Quantum Learning Systems (now Quantum Transformation Technologies). This journey typically takes place over 18–24 months (though at NYGH they continued from 2003 through 2007) and helps the organization build capacity in systems thinking and organizational learning through coaching, workshops with the Executive Team, and parallel workshops with the Management and Leadership Team (consisting of 150 Directors). In addition, the CEO, Board, and Executive Team built a Balanced Governance Scorecard and cascaded the resulting metrics down through the organization. The outcomes of the Quantum Journey are measured by several surveys that assess the organization's Human Capital Outcomes at regular intervals. These surveys can be considered as proxies for climate surveys in safety culture, but in order to make this analogy we must first consider the Human Capital Outcomes work on its own terms. The Human Capital Outcomes measured by the Quantum-NYGH Team included eight areas:

- Strategic Competencies.
- Personal Growth and Leadership Development.
- Culture Shift.
- Accountability.
- Personal and Organizational Learning.
- Strategic Alignment.
- Leadership/Management Balance.
- Strategic Integration.

The first five of these were measured by periodic surveys; the last three were evaluated more qualitatively in relationship to critical incidents and ongoing discussion at the Board level, by the CEO, and within Administration. The framework for these discussions was the Balanced Governance Scorecard and corporate dashboard (to be reviewed in the next section).

Strategic competencies were defined as: 'Everyone in management shares a common language/ framework for identifying leverage and achieving the outcomes set out in our Balanced Scorecard.' Tools and models included in this competency were as follows:

- The Systems Leverage Model.
- Surfacing and Testing Assumptions.
- Reality/Vision Gap.
- Dialogue/Discussion/Reflection.
- Balanced Scorecard/Themes.
- Framing and Reframing.
- Story-Telling.

Survey results indicated that the organization went from 95 per cent 'poor' to 'fair' on strategic competencies, to 88 per cent 'good' to 'very good' at being strategic.

Personal growth and leadership development competencies were defined as: 'Everyone in management has a deep understanding of themselves. People understand what they need to learn to grow as a leader as the hospital transforms.' Tools and models included in this competency were as follows:

- Personalysis Report.
- Personalysis Owners' Manual.

- Generative Coaching.
- Koestenbaum 360° Feedback.
- Learning Styles Survey.
- Leadership Reflection.
- Personal Vision.

Survey results indicated that the organization went from 82 per cent 'fair' to 'good' at personal growth, to 92 per cent saying 'good' to 'excellent' in their leadership development.

Cultural shift competencies were defined as: 'Managers can lead processes which internalize our emerging shared vision, create rituals that enable people to practice and live by the hospital's values, and align our thinking and behaviour to our emerging strategy.' Tools and models included in this competency were as follows:

- Culture Shift Surveys.
- Surfacing Undiscussables/Conflict Resolution.
- Learning How To Learn.
- Rules-of-the-Road/Team Learning.
- Personal Change Strategy.
- Personalysis/Talent Management.
- Generative Coaching and Accountabilities Dialogues.

Survey results indicated that the organization went from 95 per cent 'poor' to 'fair' on culture shift capacity, to 78 per cent 'good' to 'very good' at shifting culture.

Accountability competencies were defined as: 'Every manager has an *Accountability Agreement* which sets out specific outcomes/measures/targets which they are accountable for achieving in the *Organizational Balanced Scorecard* – as well as the supports required to achieve these; the positive and negative consequences, and, their *Personal Learning Contract* that will enable them to be successful.' Tools and models included in this competency were as follows:

- Accountability Agreement Tool/ Process and Talent Management.
- Setting Indicators/Targets.
- Stretch Goals.
- Managing Up/Time Management.
- Feedback/Coaching.
- Personal Learning Style.
- Personalysis.
- Becoming a Leveraged Person.
- Appreciative Inquiry.

Survey results indicated that the organization went from 82 per cent 'poor' to 'fair' on accountability, to 63 per cent 'fair' to 'good' on accountability.

Finally, *Personal and Organizational Learning Capacity competencies* were defined as: 'Managers can utilize action learning methods/tools to accelerate group and personal learning outcomes.' Tools and models included in this competency were as follows:

- Action Learning/Story-Telling.
- The Hourglass Model.
- The Story Wheel.
- The Landscape Map.
- The Situation Map.
- Pattern Analysis.
- Collective Intelligence.
- Systems Thinking/Team Learning.

Survey results indicated that the organization went from 95 per cent 'poor' to 'fair' on personal and organizational learning to 68 per cent 'good' to 'very good' on learning.

In summary, NYGH showed consistent gains in these five Human Capital competency areas in conjunction with the capacity-building programme, from Poor to Fair outcomes to Good, Very Good, and sometimes Excellent outcomes. The other three Human Capital Outcomes – Strategic Alignment, Leadership/Management Balance, and Strategic Integration – were evaluated by the CEO, Senior Staff, and Board in terms of overall hospital objectives and outcomes, as defined by the Balanced Governance Scorecard and Corporate Dashboard.

Safety Strategies for Transforming Mission, Leadership, and Culture

The next layer of the Safety Culture Pyramid includes organizational mission, leadership strategies, norms, history, legends and heroes. In this section we review strategies from organizational transformation which NYGH has pursued over the last several years.

In the post-SARS era, Bonnie Adamson (CEO), the Board, and the Senior Staff used several key strategies to revive and invigorate the organization. Aside from the Quantum Transformation Journey, these included the following:

- Consistent attention to mission, vision, and values.
- Use of a Balanced Governance Scorecard and Corporate Dashboard to align, guide, and monitor performance.
- Consistent, active dialogue and engagement with all levels of the organization, and especially with the Leadership and Management Team.
- Active formulation and execution of innovative strategies in quality, risk, and patient safety.

Mission, vision, and values North York General's mission, vision, and goals are as follows: Mission: 'A community teaching hospital in a continuum of healthcare, providing compassionate and quality care to diverse communities in North Toronto and beyond.'

Evolving Vision: Community of Success: Serving with Kindness:

- Each role is essential.
- A well designed, safe workplace makes it easy to do the right things right.
- System relationships achieve improved care for populations, patients and their families.
- Everyone is a leader in achieving quality outcomes and in leveraging resources.
- People celebrate with others the joy and success of their work.

Core Values:

- *Listening* to appreciate diversity.
- *Learning* through dialogue and reflection.
- *Leading* with courage, transparency and forgiveness.
- *Serving* patients, families and others with kindness.

It is important to note that Bonnie Adamson made several important speeches and public statements linking patient safety improvement to the overall strategy and direction of the hospital. With the help of Quantum, she articulated the connection between the transformation journey, a culture of commitment and caring, and patient safety outcomes. This clear linkage of the vision and mission with patient safety became one of the hallmarks of Bonnie's leadership as CEO, and is a theme which she is studying as of this writing.

Use of a balanced governance scorecard and corporate dashboard to align, guide, and monitor performance Drawing on the work of Kaplan and Norton, and guided by the Quantum team, NYGH developed and used a Balanced Governance Scorecard to organize and monitor organizational performance. This approach uses a set of quadrants, including Financial, Customer, Process, and Human Capital Outcomes. NYGH's current (2009) Scorecard has the following performance categories:

- Operational and Clinical Excellence.
- Knowledge Generation and Translation.
- Community Integration and System Priorities.

Operational and Clinical Excellence has two sub-categories: 'Patient and Family Experience' and 'Quality and Safety.' Within each performance category, the organization looks at desired outcomes, required processes, and enablers.

The methodology also calls for cross-cutting strategic themes, each of which is advanced by a theme team led by an executive sponsor. NYGH's current themes are:

- Patient and Family Experience.
- Quality and Safety.
- Knowledge Generation and Translation.
- Community Integration and System Priorities.
- Organizational and Human Capital.

Given our focus, let us turn next to the ways in which Quality and Safety are located within the Corporate Dashboard and Strategy Management System. Within the Corporate Dashboard, Quality and Safety are grouped together under 'Operational and Clinical Excellence.' Nine indicators are highlighted:

- Hospital Standardized Mortality Ratio (HSMR).
- Clostridium Difficile Toxin (C. Diff.) Associated Disease (Incident Rate per 100 Patient Days).
- Vancomycin-Resistant Enterococcal (BRE) Bacteremia (Incident Rate per 100 Patient Days).
- Ventilator-Acquired Pneumonia (Incident Rate per 100 Ventilator Days).
- Central line infection (Incident Rate per 100 Central Line Days).
- Surgical Site Infection-Primary Hip and Knee Surgery Patients (per cent of Patients with Antibiotic Administration within the Appropriate Time Prior to Surgery).
- Hand Hygiene Compliance (per cent compliance before and after contact)
- Skin Ulcer Prevalence.

It appears that these indicators were identified through a combination of sources, including:

- NYGH's work with IHI (e.g. ventilator-acquired pneumonia and central line infection rates were two of the six priorities for the 100,000 Lives Campaign).
- Provincial priorities (there have been significant and well-publicized outbreaks of C. Diff. in Quebec and Ontario in recent years).
- Global industry standards, such as HSMR and Hand Hygiene Compliance.
- Internal priority-setting by the staff, in consultation with Quality and Risk Management and the Board.

Consistent, active dialogue and engagement with all levels of the organization, and especially with the leadership and management team Finally, at NYGH, the third layer of the Safety Culture Pyramid (organizational mission, leadership strategies, norms, history, legends and heroes) has been heavily influenced by

Bonnie Adamson's leadership style and by ongoing dialogue with all levels of the organization. This has occurred through the Quantum Journey, through ongoing meetings of the Leadership and Management Team, and through various public speaking and discussion forums in which Bonnie has played a key role, including Board meetings, the Ontario Hospital Association, leadership and management workshops, and Rose Garden talks, in which Bonnie addresses a topic at an open-invitation event for all staff. Bonnie has also been a teacher-leader, in that she often leads the management workshops herself, with help from internal and external organizational development resources.

One example of Bonnie's dialogue-based leadership is a Rose Garden session. Though not literally in the Rose Garden, these sessions provided an opportunity for Bonnie to engage with staff in the work on story-telling and story-sharing which the author had introduced to the organization. She wanted to try out the method for herself, and also to engage staff in it. With some background coaching, Bonnie led two sessions in which she asked staff to share their stories.

During these dialogues, Bonnie discovered several examples of staff using their stories and intuitions to improve decision-making. For example, consider the following:

- A nurse was under pressure to discharge a patient from the ICU, in order to maintain patient flow. Observing him, she had a feeling something wasn't right; she said 'Don't discharge him'. Two hours later, the patient went into cardiac arrest; the team was able to save his life.
- During a management and leadership dialogue on ED/GIM, it became evident that when there are bottlenecks in patient flow, sometimes simple things can help the most. Two weeks later, housekeeping recognized there was a bottleneck and pitched in, working overtime, until it was resolved. No one asked them to do this; they noticed and responded to the situation, based on their new understanding and situational awareness.

These are exceptions to normal rules and protocols. They may be rare, but they indicate a growing ability of staff to reduce variation in practice and, at the same time, to use judgment and healthy team interactions to make good, balanced decisions on behalf of patients.

Cultural Transformation: History, Heroes, and Drivers

There is one more aspect of the third layer of the Safety Culture Pyramid (organizational mission, leadership strategies, norms, history, legends and heroes) which deserves closer attention: 'norms, history, legends, and heroes'. This layer is largely invisible from within an organization, but it can be seen more clearly from the researcher's outside perspective.

One part of the North York story since 2003 involves a kind of 'rising from the ashes.' From tragedy and trauma grew opportunity, to reshape and re-form the organization. Bonnie Adamson took on this task by pursuing what she calls the 'Culture-Leadership Transformation'; she was supported and helped by the Board, led during most of this era by Gordon Cheesebrogh, and by external help from the Quantum group. In the Quality and safety area, the organization's revival was supported by Susan Kwolek and resulted in North York's growing reputation, increasing approval and attention from the Ministry, and a number of resulting innovation grants. From an external view, it appears that leaders such as Bonnie, Gordon, and Sue became implicit 'heroes' and set a tone, and standard, for the entire organization that created a culture of innovation. This culture was geared toward trying new things toward fuller realization of the mission.

The more visible manifestation of this culture, and of the NYGH patient safety journey, is contained in the *Annual Quality Report*. Since 2008, this report has included Patient Safety Drivers, such as:

- Reduce Mortality.
- Reduce Infection Rates.
- Reduce Adverse Events.
- Enhance Culture of Patient Safety.
- Improve Communication.

For each of these, there are subset factors. For instance, for 'Enhance Culture of Patient Safety,' these include the following:

- Patient Safety Culture survey.
- Just Culture.
- Walkabouts.
- Disclosure.
- High Reliability Organizations.

For Quality, the Drivers are broader:

- Improve Effectiveness, Accessibility, and Efficiency.
- Enhance Continuity of Services and Client-Centreed Services.
- Worklife initiative (support wellness in the workplace).

For our purposes, we will focus on the Safety drivers. How did they emerge, take shape, and evolve over time? What was the 'story behind the numbers' achieved by the Lean-Kaizen efforts, reflected in the Human Capital Outcome Surveys, and correlated (it appears) with the dramatic reduction in Ventilator-Assisted Pneumonia and Central Line Infections? Does that story reflect the culture of innovation noted earlier?

Rhonda Schwartz, the Director of Quality and Patient Safety, recalls:

When I came in the summer of 2004, the organization was already working on Patient Flow and Central Line infections. QURUM (the quality and risk management group) was pulling people together. I recall asking, what do we need for Quality? In 2004, IHI started a Forum on reducing variation in practice by implementing six known best practices in acute care. They said they were limiting this to US hospitals; we got going on our own. In December 2005, IHI announced the 100,000 Lives Campaign. This was promoted in Canada as Safer Healthcare Now; we did all six projects.

In 2006–7, we hired a Lean Six Sigma expert and two engineers to help us with lean process redesign. The Ministry of Health came to us in 2006 and asked us to be a pilot site; we used their funding to hire Simpler, an outside consulting firm to help with Lean-Kaizen events.

Throughout their Lean journey (which continues to date), NYGH has been committed to retaining the knowledge and practice contributed by others (such as Simpler), consistent with their commitment to capacity-building. The hospital worked to integrate outside help with the preexisting culture and improvement methods, which emphasized dialogue and culture-leadership transformation. The hospital leadership recognized that adapting the Lean-Kaizen methods to their culture was a key to success.

In 2008, QURUM initiated Six Sigma Projects designed to 'look deeper.' These were six-to-nine month projects, designed to dig deeper with data. The project steps were to Define, Analyze, and Measure, for instance with hospitalists. A key issue was metrics. Rhonda Shwartz says, 'For two years, we weren't sure where we wanted to be' (regarding metrics). One outgrowth of this stream of activity in quality and safety was that a number of process improvements were 'baked in to Critical Care', in such areas as:

- Medication Reconciliation.
- Acute Myocardial Infarction.
- Ventilator-Acquired Pneumonia.
- Central Line Infections.
- Surgical-site infections (through a Surgical Safety Checklist).

These improvements have been organized around key organizing principles and methods, including cross-area engagement, the IHI 'bundle' concept, and Lean/ Kaizen language and activities. In addition, they have been linked back to Strategy through the Operations Excellence theme team and indicators, with particular emphasis on Quality, Safety, and Patient Experience. In this regard, front-line process improvements have been linked to operational methods and approaches, and ultimately back to overarching hospital strategy. In brief, the NYGH culture of innovation has enabled widened impact and visibility for patient safety.

Incident Review for 'Underlying Beliefs and Unspoken Assumptions'

The purpose of case analysis is to establish retrospective findings that can be used to identify areas of strength, weakness, and potential improvement. In the Patient Safety field, critical incident review is a best-practice method of case analysis. North York General Hospital has used critical incident review at the Board, administrative, and staff levels. This section will review that work as an input to an overall assessment of safety culture and its recent evolution at the hospital.

Starting in 2005, the hospital administration began sharing critical incident reviews (of adverse events) with their Board Quality Committee. This effort was led by Bonnie Adamson (CEO), Susan Kwolek (Sr. VP of Quality), staff members from Quality and Risk Management, and the respective Board Chairs, Dunbar Russell and Bruce Rothney. The purpose was to more fully engage the Board with the staff in improvement efforts by bringing the 'bedside' into the 'Boardroom'.

Incidents chosen for review were based on Quality priorities, as defined by Sue Kwolek, reported in the meeting by invited staff, and discussed by the Committee. Implications and potential lessons were identified, and a shorter three-month follow-up was often conducted.

The benefits of this review process were evident from the outset. To start with, the Board had an awakening about the realities of life in a hospital; at first, this came as a shock and surprise: 'you mean things like that actually happen in the hospital?'. At times it triggered anxiety, blame, and command-and-control behaviour, as in demands for quick fixes in policy, procedure, or rules. But the Board members could see that Bonnie and Sue were serious about partnering with them, and they began to appreciate that the 'fixes' were more complex than they had thought at first. They began to appreciate the challenges faced by staff and responded to Sue's insistence on following 'Just Culture' principles.

Storytelling – managing flex beds when emergency is full The storytelling methodology was introduced to the Leadership and Management Team at North York General in 2006 by Tom Bigda-Peyton. This methodology uses story, narrative, and anecdotes to look at observed practice in a given area, to discern the logic of action being used by individuals, groups, and the organization as a whole, and to infer potential solutions to chronic problems and compelling opportunities that would otherwise not be visible or accessible.

Our work on safety culture has led us to believe that healthcare delivery environments need early warning mechanisms and preventive alerts for instance about the unintended consequences of changes in the system. For instance, when a change happens in the clinical environment due to new policies, procedures, or resource constraints, it is usually not classified as a 'safety' or 'risk' issue yet may represent an unrecognized 'accident waiting to happen'. We have found a persistent need in these environments to elicit information about deviations so as to access and evaluate information regarding underlying sources of error, make meaning, and look for patterns and underlying failure sources.

From a cultural standpoint, we know that stories create (and maintain) culture. How do we shape what people attend to and what they believe they are accountable for? How do we move from current stories to new stories? How do we move from discipline-specific stories of individual heroics under crisis to stories of shared accountability which include individual excellence?

At North York General Hospital, we accomplished all of these objectives with a Qualitative Analysis tool – the Story Map™. In 2006, Tom Bigda-Peyton facilitated a demonstration of the story mapping process with the Leadership and Management Team (including 150 managers). This turned into an important recalibration point for NYGH. It also shows the linkage, in action, between the different levels of the Safety Culture Pyramid. The tool (and associated facilitation process) also enabled the organization to achieve the following:

- Make the connection between patient safety and other areas of performance.
- Provide an early warning mechanism by identifying unintended consequences of current practices and responses to changes in the system.
- Start moving from a culture of blame and individual heroism to one of shared accountability with individual excellence.

Using the story map   The sense of concern, and even urgency, about the 'flow' problem was widely shared. Indeed, one of the most powerful standouts was a collective 'aha' moment which happened when Tom Bigda-Peyton asked the group, before proceeding, how many of them were affected by this problem. Easily 75–80 per cent of the hands in the room went up; in itself, this seemed to be a surprise to the group. Later on, Susan Kwalik (Senior VP of Quality) referred to this as a turning-point, a positive sentinel moment. From that moment on in the meeting, she felt, there was agreement that 'we're all in this together'. This standout does not show up on the map, but it was sparked by the exercise of working through the map.

A project team had been convened to look at policies, and potential guidelines, for the 'use of flex beds when Emerge [ncy] is full'. Using the Story Analysis™ method, it quickly became apparent that this statement of the problem was not useful, because (as the group confirmed) 'Emerge is *always* full.' Thus we changed the focus of the inquiry, early on, to the guidelines for the use of flex beds '*when there is gridlock*'. (The term 'gridlock' came up as a useful description, or metaphor, for the situation that the group could recognize and use in action.) We concluded that, for the most part, the use of flex beds helps in such situations. Thus, our first 'finding' was a revised assumption: 'When in gridlock, flex up.'

We had to go farther, though, because of the wide variation in practice, across units, on the use of flex beds. There was a zone of common practice (in terms of numbers of beds) between several units, Telemetry, and Medicine. Obstetrics had no established practice, Surgery and Mental Health were using different guidelines, and Paediatrics had well-developed rules of thumb which weighed and balanced four or five factors in decision-making. No one was aware of these differences before, and many departments (such as Housekeeping) had no knowledge of how others were responding to the situation or how they could assist. In effect, each local unit was coming up with its own solution to the problem, independent of others.

This situation, which was previously recognized only as a frustrating problem of changing policies, procedures, and guidelines, was creating a variety of unintended outcomes. For one thing, there was no recognition of how the efforts of local units were making a difference. The problem could not be solved by a unit or departmental perspective alone, but local units could look at the map and see the benefits of their solutions. They could be recognized (albeit implicitly) for their contributions, learn from other units, and begin seeing synergistic options for next-generation solutions to the overall problem.

While the use of flex beds was working to some extent, the overall situation still yielded very significant unintended consequences, many of which directly impacted patient safety. In fact, one key aspect of the map is that it demonstrated that patient flow was an important patient safety issue. As the sense of 'gridlock' increased across clinical areas, frustration rose among staff and patients, 'acting-out' by patients increased, and there was a growing risk of adverse events which would not have happened otherwise.

During the discussion, it became clear that using flex beds was only one of several potential high-impact strategies for dealing with the problem. As a whole group, we began to explore a variety of strategic, operational, and tactical options for approaching the problem. These ranged from adjusting the practice for bed alerts, to coordinating communications across units, to examining interconnections with the Province on after-care. For our purposes, the most important options are strategic. Although they are the hardest to implement, they also have the greatest potential for sustainable, systemic impact.

When we look for the major leverage points to shift the picture from 'gridlock' to a smooth 'flow' of patients through the hospital, the areas of greatest impact have little to do with specific actions, such as revised policies, procedures, structures, and processes. Instead, *the areas of leverage lie in the operating assumptions* that are held as obvious, and which can only be surfaced through this kind of story-mapping.

The key operating assumptions in the NYGH Flow Map are as follows:

- Doctors can only release patients when they are 100 per cent ready.
- Emerge is full all the time – it operates at 97 per cent capacity.
- The hospital operates on a nine to five, five-day schedule (while patients operate on a seven day, 24 hours a day schedule).

These assumptions are deeply embedded in the organization's habits, practices, structures, policies, and procedures. They are hard to change because they are so taken-for-granted and because they are reinforced every day by standard operating procedures. Yet changing them represents the best chance for breakthrough on a difficult, seemingly intractable problem.

In short, and as noted earlier, the use of the story mapping process enabled the organization to achieve the following:

- Make the connection between patient safety and other areas of performance (such as patient flow).
- Provide an early warning mechanism by identifying unintended consequences of current practices and responses to changes in the system (i.e. when there was 'gridlock').
- Start moving from a culture of blame and individual heroism to one of shared accountability with individual excellence (i.e. 'we're all in this together').

In addition, by surfacing underlying beliefs and assumptions, we created an opportunity for the organization to reflect on its own paradigm and make appropriate adjustments.

Using the Safety Culture Pyramid

We can see that behaviours are influenced by 'Attitudes', 'Mission/Leadership/ etc.', and 'Underlying Assumptions'. That said, these layers of the pyramid live in different spaces or zones of activity; they are related but also somewhat independent of one another. Attitudes and Opinions, as measured by surveys, are closer to what Argyris and Schön have called 'espoused theory': that is, how we think we would behave in certain situations, or how good we believe we are at something. Underlying Assumptions are closer to the area that Argyris and Schön have called 'theory-in-use', or how we behave under pressure in the 'moment of truth'. The layer of Mission, Leadership, etc. lies in a middle zone between what is espoused and what is lived, or practiced, at 'moments of truth'. As we move down the layers, we can see the factors that predict behaviour in action at the moment of truth. This is a key benchmark of safety culture-in-action, or the safety default position, that defines individual and organizational behaviour under pressure.

Thus we find in looking at the story map that under pressure (e.g. 'gridlock') NYGH staff reverted to normal, unit-based assumptions to drive behaviour (e.g. we are here from 9 a.m. to 5 p.m.; doctors can only release patients when 100 per cent ready). These Underlying Assumptions drove hazards and potential patient safety issues, but until they were surfaced and examined (using the concept of Unintended Consequences for the whole system, and the whole problem) they could not be adjusted, much less altered.

The Evolution of Safety Culture at NYGH, 2004–2009

The next phase of safety culture evolution at North York is making it a way of life. In thinking about High-Reliability and Ultrasafe Organizations, Rhonda Schwartz says, 'people don't know the principles, the mindset'. This implies a need to articulate, teach, and learn the principles of highly reliable organizations. It means moving from 'projects' and 'events' to 'it's an ongoing way of life'. And it means cultivating a new mindset; in other words, changing our thinking to reflect the new assumptions and underlying values of highly reliable and ultrasafe organizations.

One example of this shift might come from organization-wide internalization and practice of principles of highly reliable organizations, such as 'situational awareness', 'reluctance to simplify', and 'preoccupation with failure'. Situational awareness refers to the ability of individuals on a care-giving team, and also of the entire team, to perform their work with a two-track mindset: one track is performing their own part of the job; the other is maintaining a sense of the overall progression of the work, and adjusting their part accordingly if something has gone wrong or needs just-in-time revision. Another such principle is 'reluctance to simplify'. This involves following rules and procedures while also looking for important exceptions; in medical care, it might mean looking for a secondary diagnosis rather than remaining satisfied with the initial finding. A third principle of high-reliability organizations and teams is 'preoccupation with failure'. This means recognizing that things can and will go wrong, that complacency (even in well-performing organizations) is a patient safety hazard. Accordingly, organizations that are preoccupied with failure do continuous review and scenario planning; they design contingencies and safeguards on the assumption that things will go wrong, not on the hope that they won't. All three of these principles require a shift from typical assumptions and thinking in healthcare; none can be satisfied entirely by projects and events.

In effect, this means the internalization of a 'Reporting' and 'Just' culture as the normal, everyday, operating practice. What might this look like? North York General is considering an initiative to build organizational resilience. This is the capacity for an organization to 'bounce back' from a hard hit, as North York did after SARS; but it is also the institutionalization of that capacity, for instance by reinforcing it with systems, structures, and processes. In practice, this could include the following dimensions:

- At the *staff* level, asking questions such as: What could go wrong? What would you do if it happens?
- At the *operations* level, conducting both after-action and prior action reviews.
- At the *strategic* level, engaging in ongoing scenario planning and learning, in ways that build on emerging signals from the environment as well as the work at the tactical and operational levels of the organization.

Future Directions

There are two interesting areas that hold great promise for future research: Resilience and Crisis-to-Strategy shift. In this section we present an introductory discussion on these two topics and encourage the readers to explore these topics further.

Resilience

Many of the themes pursued in this book have emphasized the need to do assessment, intervention, and evaluation toward sustained improvements in Safety Culture. In effect, we have posited a maturity model of Safety Culture, with accompanying critical success factors (behaviour, climate, strategies, and underlying values). We have also proposed a way of evaluating, or measuring, movement toward a healthier safety culture, by defining two scales (accountability and learning) and four stages of maturity (Secretive, Blame, Reporting, and Just Cultures). We have presented case studies that illustrate these dimensions of safety culture and described approaches used by specific organizations to attain higher levels of safety culture maturity. Thus far, however, we have left out an important element: resilience. In this section, we introduce this concept, use it to integrate our argument thus far, and point toward ways of incorporating it in subsequent research and development that can link operational improvements and academic research in the journey toward system-level safety improvement which can be sustained over time.

Resilience is a core capability of well-performing organizations in aviation, healthcare, and other high-consequence industries. Leaders need to learn how to help their organizations practice resilience. A 'resilient' organization, team, or individual has the ability to recognize when day to day work is operating at the margins of expected performance and safety, and is able to adjust practices to return the organization, unit or a specific patient/customer interaction to a state of balance or optimum level of functioning. In high-consequence industries (aviation, chemical, nuclear, and healthcare), it involves not only the ability to recover from accidents, but also the possession of other key attributes, such as vigilance, that can prevent harm before it happens. In organizational terms, it means the capacity of a government, healthcare system, or private-sector company to recover from environmental shocks, adapt to the new situation, and return to a well-performing equilibrium point.

As noted already, organizations operate within a complex myriad of competing priorities, goals and constraints. Resilience focuses on how to help people manage complexity under pressure and continue to achieve success. A resilient organization will look for ways to understand the shifting environment in which it operates and will learn to recognize vulnerabilities and hazards so as to defend against potential failures and create pathways to success. Brittleness is the antithesis of resilience.

Strategic risk-taking and risk management, toward optimizing performance, are hallmarks of the Resilient Organization perspective:

> Well-performing organizations encourage risk taking. They are willing to try new methods when common sense dictates that better results can be achieved by following the spirit of a regulation, instead of the letter. However, staff must hold the values of stewardship, service and results, and they must consult with each other. When their people are governed by these values, the well-performing organizations encourage risk taking as a matter of strategy (Otto Brodtrick, 1988 Report for the Office of the Auditor General, Government of Canada, 'The Well-Performing Organization').

Building organizational resilience is a necessary performance platform in today's environment. Sustaining results and reliability requires the ability to respond to surprise, adapt to changing conditions, and anticipate possible trends before they play out. Leading for resilience helps to identify, nurture, and develop these capacities in large, complex organizations.

Resilience in organizations can be seen in various ways:

- As a naturally occurring characteristic of organizations which have matured their safety culture. Organizations and groups within them can discover that they have become more resilient in the face of budget crises, leadership changes, or other 'shocks to the system' when their culture is more mature.
- Resilience engineering, as in the work of Hollnagel, Woods, and Leveson (2006). In this sense, the principles of a resilient organization (e.g. failsafe engineering) are used for organization design and development. Methods include building the capacity to live up to the four attributes (to Learn, Respond, Monitor, and Anticipate), manage the efficiency-thoroughness tradeoff, and look at good things that happen as well as bad.
- As a proxy and stand-in for organizational learning in the Argyris-Schön-Bateson tradition (of single, double, and triple-loop learning); this involves whole system learning-in-action, making decisions under uncertainty, and navigating through choppy, turbulent waters toward a new destination and future.

These ideas are consistent with everything we have said, but they give us new directions to pursue and capabilities to build as we follow up on the implications of the argument thus far. In brief, as an organization's safety culture matures, the need for resilience also grows. Otherwise, the organization will 'fall back' at the first sign of trouble and will fail to recover from shocks to the system. On a day-to-day basis, it will have trouble rebounding from smaller 'bumps in the road' and helping people to regroup, stay on track, and form the ability to more effectively plan for, and anticipate, a future which is both predictable and highly uncertain. Living with this paradox, which is embodied in part by the

Efficiency-Thoroughness tradeoff, is a hallmark of a resilient organization. After-action review, in the sense of reviewing and learning from incidents, is another. And the ability to manage dilemmas, paradoxes, and tradeoffs as a normal part of worklife becomes a capacity to be fostered and brought into balance with more linear, familiar routines.

We have seen instances of these capabilities in the book, but we have spent less time on how they come into play in mature safety cultures (such as aviation, chemical, and nuclear). Rather, we have chosen to focus on the more behavioural and cultural dimensions of the problem, and also to look at the respective journeys of healthcare and aviation. Future paths for inquiry might explore the journeys of other high-consequence industries, less to understand how they have achieved 'perfection' but more to learn from what they did when they realized, despite all of their efforts, that perfection cannot be attained. At the moment of this realization, we need the resilience view.

Crisis to Strategy

Pearson and Mitroff (1993) point out that when managers define a crisis five dimensions usually emerge. The crisis event is typically of high magnitude, requires immediate attention, produces a feeling of surprise, demands action, and involves a perception that the event is beyond the organization's normal control. These authors describe five phases of crisis management, which include signal detection, preparation/prevention, containment/damage limitation, recovery, and organizational learning from the event (which provides feedback and influences the next round of signal detection).

The dominant state of a Safety Culture can be described in terms of a crisis response scale, which captures how an organization prepares for, understands, and responds to serious adverse events. Figure 8.8 presents four approaches to handling crises including crisis-fixing, symptomatic, root-level, and strategic preparation and management (c.f., Patankar and Sabin, 2008). The crisis response scale represents the improvement in safety culture maturity as organizations move away from a preoccupation with reactive crisis-fixing and toward a strategic, systemic, and systematic plan of action to deal with crises. Shortell and Singer (2008) argue that patient safety can be improved by such a strategic systems approach.

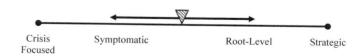

Figure 8.8   Crisis response scale

The two poles of the Crisis Response Scale represent differences between reactive, short-term responses (crisis fixing) and long-term, proactive approaches (strategic planning, prevention and management). Smallman (1996) differentiates reactive from proactive approaches to risk management. Reactive risk management is driven by the occurrence of events and offers a lagging response; it is reactionary to the course of unfolding environmental events and based on forecasting models. Reactive approaches manage risk by either accepting certain levels of loss or seeking to transfer risk and loss to another who will 'bear the cost'.

Conversely, Smallman (1996) maintains that proactive risk management seeks to avoid, reduce, and prevent risk. Rather than being pushed by events of loss it is pulled by risk avoidance. Rather than relying solely on probability theory or actuarial models, proactive risk management also incorporates qualitative techniques from organizational learning such as scenario planning to anticipate and plan for crises. Smallman maintains that a proactive and holistic approach to risk management can more effectively incorporate information from quantitative and qualitative techniques to strategically analyze and assess risks from human, organizational, technological, regulatory, infrastructural, and political sources. Likewise, Vaughn (1999) analyzed the occurrence of mistakes, misconduct and disasters in organizations and concluded that they are systematically produced by the interconnections between environmental factors, organizational structures and conduct, cognition, and choice. Understanding this interconnectedness should allow for better strategic approaches to prevent, minimize, and manage adverse events.

Event focus Pearson and Mitroff (1993) argue that crises are preceded by early warning signals that are often ignored or actively blocked. Crainer (1993) states 'disaster after disaster has revealed the same pattern of incompetence, lack of foresight and irresponsibility. Safety systems ... not updated to cover changes in technology and operating conditions. The employees who implement them have not been adequately trained' (p. xiii–xiv). An organization that operates with a secretive or blame safety culture is ill-prepared to deal with crisis. When disaster occurs the organization mobilizes its resources to fix the crisis and minimize damage and loss to the best of its ability. Secretive and blame cultures do not have the capacity to pay attention to early warning signals that indicate a crisis in the making.

As Pearson and Mitroff (1993) point out, when a crisis occurs containment strategies and procedures have to be implemented; however, these processes need to have been developed, validated, and readied for deployment before the occurrence of the actual event. Organizational recovery from crisis requires both short-term and long-term strategies to return to normal operations. Failure to think and plan ahead strategically will typically yield poor results in responding to and recovering from a crisis.

Baumard and Starbuck (2005) found that learning often does not occur following failure. They caution that 'one should be suspicious of efforts to explain failures in terms of idiosyncratic circumstances or external events' (p. 295). The unprepared may learn little from their own direct experience of failure; however, strategic thinkers may profit from reflecting on instructive case studies of failure. This echoes philosopher George Santayana's (1905–1906) famous comment on the importance of learning from others' experience: 'Those who cannot remember the past are condemned to repeat it'.

Elliott (2009) documents how organizations can fail to learn from crisis, beginning with deficiencies in the initial acquisition of knowledge, to dissemination, and institutionalization of changed norms that impact standard operating procedures and practices. Elliott, Smith and McGuinness (2000) argue that rather than being unique occurrences, disasters have similar patterns of 'incubation, latent error, managerial incompetence, a lack of foresight and, more reprehensibly a lack of hindsight' (p. 17). They detail a number of factors that prevent learning from disasters, including simplistic explanations, scapegoating to project blame, insufficient regulation, inadequate problem definition, and complacency. The authors maintain that there is a lack of empirical studies that deal with crisis learning and a corresponding danger of reductionistic explanations that prevent a systemic view of complex problems and the emergent properties of dynamic systems. Elliott, Smith, and McGuinness contend that effective levels of trust, communication and corporate responsibility need to be present if organizations are to learn from crisis.

Symptomatic Focus Common patterns of problem solving in demanding operational settings often require overcoming immediate obstacles by focusing on the presenting symptoms rather than reflecting on the underlying causes. A focus on symptoms leads to short-term solutions but does not deal with the recurring nature of problems that have not been addressed at a deeper level, and may lead to further serious consequences. This is analogous to the work of Argyris and Schön (1978) on first and second order learning. First order learning focuses on solving the immediate problem, while second order learning goes deeper to assess, understand, and prevent the reoccurrence of similar problems in the future by addressing the underlying cause. Tucker, Edmondson, and Spear (2002) found that first order problem solving was very common in patient care tasks since time was of the essence and healthcare workers focused on doing what was required at the moment. Their research found that 'a lack of available time and norms that valued quick, self sufficient solutions contributed to a pattern of front-line workers rarely engaging in root cause removal' (p. 134).

McLucas' (2000) review of accidents found a distressing failure to recognize emergent patterns of events, which resulted in recurrent failures to learn. He concludes that managers and strategic decision-makers fail to appreciate the dynamic and detailed complexity of the systems that they are responsible for managing. Smith and Elliott (2007) examined barriers that can prevent learning from crisis situations. While many would like to assume that learning takes place after negative events, it appears that there are powerful forces working against it.

In their discussion of the interactions between learning and crisis, they differentiate 'learning *for*, *as*, and *from* crisis'. Learning *for* crisis is preparatory and pre-crisis; it deals with developing the skills and abilities to manage a crisis when it occurs. Learning *as* crisis suggests the critical challenges to core assumptions and beliefs that self-analysis and reflection on a crisis event can cause for individuals in the impacted organization. 'Crisis provides the opportunity to consider the unthinkable, revealing the inadequacies of previous assumptions and practices' (p. 526). Learning *from* crisis describes how an organization can hopefully learn from its own and others' experiences in dealing with difficult events. However, many organizations often pay little attention to lessons learned from other's accidents or even their own near-miss events. While Smith and Elliott note a variety of obstacles to crisis learning, the most common barriers are rigid core beliefs, values and assumptions (which lead to misunderstanding), and ineffective communications (due to undiscussable issues, information overload and group think).

Root-level focus   Root Cause Analysis has been used in aviation, healthcare, and other high consequence industries to diagnose and correct the underlying drivers of adverse events. Such an analysis attempts to deal with the complexity of detail and the dynamic systemic interplay of factors that can combine to produce crisis. Ultimately root cause analysis attempts to eradicate the problem at its source. Wu, Lipshutz, and Pronovost (2008) concluded that when using root-cause analysis in healthcare it is often easier to identify problems than to overcome obstacles in formulating and instituting corrective actions to improve the situation.

Cannon and Edmondson (2005) offer a framework to overcome barriers to learning from failures that are embedded in technical and social systems. To overcome technical systems barriers they recommend implementing knowledge management systems that capture and organize key data while conducting carefully structured after action reviews to effectively analyze failures. To overcome social systems barriers, they recommend removing blame by implementing non-punitive error reporting systems and viewing errors as opportunities for learning (e.g. to avoid 'shooting the messenger'), and providing expertise in group dialogue to effectively address 'hot' issues. This approach can be instrumental in reframing thinking so that failure is seen as an important source of information and learning.

Tamuz (2001) addresses learning disabilities for regulators in air transportation and the challenges for organizational learning. In high-consequence industries, the threat of punishment to deter error also provides a disincentive for individuals to reveal information about significant hazards. This produces difficulties in constructing and interpreting databases to learn about potential dangers. In a review of the Aviation Safety Reporting System and the Aviation Safety Action Program, Tamuz concludes that these systems demonstrate the importance of maintaining separate safety reporting systems 'to ensure regulatory enforcement and others to promote organizational learning … and … support diverse, multiple systems for monitoring safety' (p. 298). Meyer and Eisenberg (2002) discuss the transformation of the research agenda in patient safety to a strategic integrated initiative, which may yield insights into underlying causes of medical error.

Strategic focus Pearson and Mitroff (1993) contend that effective organizations continuously and systematically seek to evaluate and mitigate risk to prevent the crisis event or manage it if a crisis should occur. Well-prepared organizations seek to learn from the crisis by reflecting on factors that helped the organization perform effectively and recognizing barriers that interfered with an appropriate response. Lessons learned need to be extracted and made part of the organization's memory for the event. However, many organizations do not review the crisis event to examine and learn from their performance, fearing that it will 'only reopen old wounds'. Pearson and Mitroff conclude that 'the purpose of crisis management is not to produce a set of plans; it is to prepare an organization to think creatively about the unthinkable so that the best possible decisions will be made in the time of crisis' (p. 59).

Bonn (2001) argues that many organizations suffer from a lack of strategic thinking. He addresses strategic thinking at both an individual level (e.g. systems thinking, creativity, and developing a vision of the future) and organizational level (e.g. holding regular strategic dialogue among leadership, and taking advantage of employees' creativity and ingenuity). He maintains that by integrating strategic thinking at these levels a critical core competency can be created that forms the basis for sustainable advantage.

Keeney (1994) maintains that strategic thinking and decision-making should be driven by core organizational values. He suggests that value-focused thinking helps recognize and identify decision opportunities, create better alternatives, and develop an enduring set of guiding principles (p. 41). Keeney and McDaniels (1992) provide an example of value-focused thinking at an electric utility company that demonstrates strategic trade-offs related to economic/business considerations, environmental impact, health and safety issues, and service quality.

Lagadec (1997) points out that preparation for common emergencies is routine (e.g. fire drills); however, organizations often do a poor job of planning for larger accidents or crises. The author suggests that senior leadership needs to make the time available to strategically plan for major unexpected events through dialogue about uncertainties and potential weaknesses, review of past difficult events, and simulations of destabilizing events to enhance learning.

Similarly, Sapriel (2003) maintains that effective crisis management requires an integrated and systemic approach that includes ongoing risk assessment, development of tested crisis response processes, regular training and practice, and a strategic approach that is supported by top leadership and implemented by managers in all key functional areas within the organization. In conclusion, as Pearson and Mitroff (1993) point out, a strategic approach allows an organization to move from crisis prone to crisis prepared.

Chapter Summary

In this chapter, we reviewed the application of the Safety Culture Pyramid Model from two perspectives: one in aviation and one in healthcare. Through these two case examples, we illustrated how different research and analysis techniques could be applied at various stages of safety culture assessment to build a comprehensive understanding of the prevailing safety culture as well as track the cultural transformation over a period of time.

As matter of future directions, we offered some basic thoughts about resilience and moving an organization from crisis-focus to strategic-focus in order to enhance the organizaton's ability to handle crises. With increasing tension between declining resources and increasing demands, organizations in high-consequence industries will need to be particularly adept at resilience without compromising their core strengths, services, or ability to manage crises.

References

Amalberti, R., Auroy, Y., Berwick, D., and Barach, P. (2005). Five system barriers to achieving ultrasafe healthcare. *Annals of Internal Medicine*, 142, 756–764.

Argyris, C. and Schön, D. (1974). *Theory in Practice: Increasing Professional Effectiveness*. San Francisco: Jossey-Bass.

Argyris, C. and Schön, D. (1978). *Organizational Learning: A Theory of Action Perspective*. Reading: Addison-Wesley Publishing.

Baumard, P. and Starbuck, W.H. (2005). Learning from failures: why it may not happen. *Long Range Planning*, 38, 281–298.

Bonn, I. (2001). Developing strategic thinking as a core competency. *Management Decision*, 39(1), 63–70.

Brodtrick, O. (1988). Well-performing organisations. In: 1988 report of the auditor general of Canada (Chapter 4). Retrieved from <http://www.oag-bvg.gc.ca/interent/English/part_oag_198811_04_e_4231.html> on May 1, 2010.

Cannon, M.D. and Edmondson, A.C. (2005). Failing to learn and learning to fail (intelligently): how great organizations put failure to work to innovate and improve. *Long Range Planning*, 38, 299–319.

Crainer, S. (1993). *Zeebrugge: Learning from Disaster: Lessons in Corporate Responsibility*. London: Herald Charitable Trust.

Elliott, D. (2009). The failure of organizational learning from crisis – a matter of life and death? *Journal of Contingencies and Crisis Management*, 17(3), 157–168.

Elliott, D., Smith, D., and McGuinness, M. (2000). Exploring the failure to learn: crises and the barriers to learning. *Review of Business*, 21(3), 17–24.

FAA (2002). Aviation Safety Action Program Advisory Circular and Guidance AC 12-66B. Washington, DC: Federal Aviation Administration. Retrieved from <http://www.faa.gov/safety/programs_initiatives/aircraft_aviation/asap/policy/> on November 14, 2005.

FSF (1998). Aviation safety: U.S. efforts to implement flight operational quality assurance programs. *Flight Safety Digest*, July–September.

Heinrich, H. (1931). *Industrial Accident Prevention: A Scientific Approach*. New York: McGraw-Hill.

Helmreich, R. and Merritt, A. (1998). *Culture at Work in Aviation and Medicine: National, Organizational and Professional Influences*. Aldershot: Ashgate.

Hollnagel, E., Woods, D., and Leveson, N. (eds) (2006). *Resilience Engineering: Concepts and Precepts*. Aldershot: Ashgate.

Keeney, R.L. (1994). Creativity in decision making with value-focused thinking. *Sloan Management Review*, Summer, 33–41.

Keeney, R.L. and McDaniels, T. (1992). Value-focused thinking about strategic decisions at BC Hydro. *Interfaces*, 22(6), 94–109.

La Porte, T. and Consolini, P. (1991). Working in practice but not in theory: theoretical challenges of high-reliability organizations. *Journal of Public Administration Research Theory*, 1, 19–47.

Lagadec, P. (1997). Learning process for crisis management in complex organizations. *Journal of Contingencies and Crisis Management*, 5(1), 24–31.

Marx, D. (1997). Moving toward 100 per cent error reporting in maintenance. In: *Proceedings of the Eleventh International Symposium on Human Factors in Aircraft Maintenance and Inspection*. Washington, DC: Federal Aviation Administration.

Marx, D. (2001). Patient safety and the 'just culture': a primer for healthcare executives. Retrieved from <http://www.mers-tm.net/support/Marx_Primer.pdf> on October 11, 2004.

McLucas, A.C. (2000). The worst failure: repeated failure to learn. 1st International Conference on Systems Thinking in Management, 426–431.

Meyer, G.S. and Eisenberg, J.M. (2002). The end of the beginning: the strategic approach to patient safety research. *Quality and Safety in Healthcare*, 11(1), 3–4.

Patankar, M.S., Bigda-Peyton, T., Sabin, E., Brown, J., and Kelly, T. (2005). A comparative review of safety cultures. Report prepared for the Federal Aviation Administration. Available from <http://hf.faa.gov>.

Patankar, M.S. and Sabin, E. (2008). Safety culture transformation in technical operations of the air traffic organization: project report and recommendations. Report prepared for the Federal Aviation Administration. Available from: <http://hf.faa.gov>.

Pearson, C.M. and Mitroff, I.I. (1993). From crisis prone to crisis prepared: a framework for crisis management. *The Executive*, 7(1), 48–59.

Perrow, C. (1984). *Normal Accidents: Living with High-risk Technologies*. Princeton: Princeton University Press.

Reason, J. (1997). *Managing the Risk of Organizational Accidents*. Aldershot: Ashgate.
Roberts, K.H. (ed.) (1993). *New Challenges to Organizations: High Reliability Understanding Organizations*. New York: Macmillan.
Santayana, G. (1905–1906). *The Life of Reason*. New York: Scribner's Sons.
Sapriel, C. (2003). Effective crisis management: tools and best practice for the new millennium. *Journal of Communication Management*, 7(4), 348–355.
Shortell, S.M. and Singer, S.J. (2008). Improving patient safety by taking systems seriously. *Journal of the American Medical Association*, 299(4), 445–447.
Smallman, C. (1996). Risk and organizational behaviour: a research model. *Disaster Prevention and Management*, 5(2), 12–26.
Smith, D. and Elliott, D. (2007). Exploring the barriers to learning from crisis: organizational learning and crisis. *Management Learning*, 38(5), 519–538.
Tamuz, M. (2001). Learning disabilities for regulators: the perils of organizational learning in the air transportation industry. *Administration and Society*, 33, 276–302.
Tucker, A., Edmondson, A., and Spear, S. (2002). When problem solving prevents organizational learning. *Journal of Organizational Change Management*, 15(2), 122–137.
Vaughn, D. (1999). The dark side of organizations: mistake, misconduct, and disaster. *Annual Review of Sociology*, 25, 271–305.
Wu, A., Lipshutz, A., and Pronovost, P. (2008). Effectiveness and efficiency of root cause analysis in medicine. *Journal of the American Medical Association*, 299(6), 685–687.

Index

accident rate 142, 182–3
accountability scale 5–6, 16, 21, 78, 113–14, 179, 191, 200
    blame culture 5–6, 49, 58, 78, 113, 166, 179–80, 182, 191, 200, 210, 229
    just culture 5–7, 17, 19, 23, 51, 58, 61, 76, 80, 93, 98, 108–9, 113, 179–80, 185
    reporting culture 5–6, 49, 58, 78–9, 93, 108, 113, 166, 191, 200–201, 204, 206, 210
    secretive culture 5–6, 58, 113, 191, 198, 200
acculturation 148, 166, 168–70, 175–8, 192
    degree of 168–9
    strategies 168–71, 175
Adverse Drug Event (ADE) 8, 62, 67, 76–7, 80, 82, 125, 140
adverse events 36, 59–61, 63, 69–70, 78–9, 82, 178, 210–11, 219, 221, 223, 229, 231
After Action Review 12–13
Agency for Healthcare Research and Quality (AHRQ) 18, 22, 25, 28, 54, 67, 78–80, 82, 121, 124–5
AHRQ Hospital Survey on Patient Safety Culture 37
Air Line Pilots Association (ALPA) 121–2, 137, 145
Air Traffic Organization (ATO) 23, 85, 87, 93, 104, 107–9, 181, 183, 191, 195–6, 201, 203–6, 234
American Airlines 73, 169
American Congress for Obstetricians and Gynecologists (ACOG) 33
American Medical Association (AMA) 24, 53, 59, 79–83, 235
anticoagulants 67–8, 77
apology 47, 71, 81

artefacts 26, 43–4, 49–50, 92, 117, 192
Ascension Health 76–7, 79–80, 139–42, 144–5, 180, 183, 185
assessment 16, 39, 41, 61, 131–2, 151, 154, 156–7, 192–3, 210, 226
assimilation 166, 168–9, 171
assumptions 9–11, 14, 45–6, 50, 92–3, 148, 150, 155, 157, 161, 173–4, 204–5, 223–5, 231
    operating 151, 166, 223
attitudes 3, 5, 12, 35, 39, 41, 45, 47, 49, 53, 95–6, 128–9, 171–2, 204–5, 224
    employee 2–3, 5, 25, 34, 108, 193, 203
Automated Dispensing Cabinet (ADC) 67
autonomy 69, 139, 149, 153, 170
aviation maintenance 42, 53–4, 82, 109–10, 129, 145, 178
Aviation Rulemaking Advisory Committees (ARACs) 137
Aviation Safety Action Program (ASAP) 7, 44, 78, 93, 118, 123, 128, 137, 143, 182, 186, 196, 201, 203–4, 231
    programmes 118, 123, 138, 182, 186, 203–4
    reports 123, 201, 203
    safety culture briefings 204
Aviation Safety Human Reliability Analysis Method (ASHRAM) 25
Aviation Safety Information Analysis and Sharing (ASIAS) 77, 143
Aviation Safety Reporting System (ASRS) 44, 137, 231

Balanced Governance Scorecard 211–13, 216
barriers 9–11, 44, 203–4, 231–3, 235
behavioural intent 38
behaviours
    organizational 166, 224, 235
    risk-taking 3–4, 7, 95

underlying 76
underlying 50, 189
Best Practices in Safety Culture Transformation 182
blame culture 5–6, 49, 58, 78, 113, 166, 179–80, 182, 191, 200, 210, 229
Board Quality Committee 210, 212, 221

Canadian Patient Safety Institute (CPSI) 212
case analysis 25, 27, 31, 191, 221
causal factors 27, 58, 62, 67, 129, 143, 191
Cedars-Sinai 78, 117, 119–21, 126
change 10, 26, 38, 42, 48–50, 90, 100–101, 110, 113, 131–2, 166–7, 171–2, 179–87, 192–4, 196–7
  behavioural 13, 129
  culture 72, 196
  programme 13, 40, 92, 97, 181
checklists 17, 45, 159
Chief Executive Officer (CEO) 4, 97, 118, 141–2, 162–5, 184, 193, 211–13, 215–16, 221
Cockpit/Crew Resource Management (CRM) 15, 17, 52, 80, 181
Comair accident 65, 74, 78, 114
Commercial Aviation Safety Team (CAST) 77, 103, 127–8
commitment
  organizational 120, 193
  statement of 208–9
compliance-based strategies 113
contributory factors 25, 31, 191
CR-BSIs 72–3
crisis 94, 99–100, 117, 120, 222, 228–35
  learning 230–31
  management 228, 232–4
crisis response scale 228–9
  crisis event 228, 231–2
  crisis learning 230–31
  crisis management 228, 232–4
cultural change 49, 62, 86, 147, 161, 179–81, 189–90, 205
cultural transformation 48–9, 147, 182, 218, 233
culture
  accountability scale 5–6, 16, 21, 78, 113–14, 179, 191, 200

blame 5–6, 49, 58, 78, 113, 166, 179–80, 182, 191, 200, 210, 229
  crisis response scale
  healthcare 54, 62
  just 5–7, 17, 19, 23, 51, 58, 61, 76, 80, 93, 98, 108–9, 113, 179–80, 185
  learning culture 23, 52
  learning scale 5–8, 16, 21, 113, 179, 191
  national 39, 147
  organization's 46, 50
  reporting 5–6, 49, 58, 78–9, 93, 108, 113, 166, 191, 200–201, 204, 206, 210
  secretive 5–6, 58, 113, 191, 198, 200

decision-making shortcuts 155–7
deculturation 166, 168
Defense Department's Patient Safety Program 136
Dental diagrams 28–9
destination statement II 140–41, 180
dialogue 7, 12–14, 26, 39, 42, 45–7, 50–53, 55, 113, 139, 147–8, 175–6, 186–7, 192, 203
DIEP flap procedure 157–9, 161–2
Dimensions of the Learning Organization Questionnaire (DLOQ) 14, 23, 39, 52
disclosure 47, 123–4, 219
disclosure of medical errors 52, 71, 80
divergence 31, 33

Efficiency-Thoroughness Trade-off (ETTO) 155, 177
Electronic Medical Records (EMR) 163
employee involvement 41, 91, 96
employee-management trust 6–7, 19, 87
employee safety attitudes 34
employee satisfaction 101, 103, 105, 199
employees
  airline 133
  frontline 102, 115, 197
  new 171, 192
error
  error management 198, 200
  non-punitive 19, 180–81, 193, 195, 200–201, 205–6

reporting system 7, 19, 39, 74–5, 131, 142, 180–81, 190, 193, 195, 198, 200–201, 205–6, 231
  see also non-punitive error-reporting system
error-producing conditions 126–7
error trajectories 58, 79, 126
errors
  self-perceived 70
  unintentional 133, 203–4
Event Review Committee (ERC) 201–3

FAA's safety culture 87, 197
factors
  contributing 3, 30, 61–2, 202
  outcome 92, 101, 106, 196, 199
failure
  active 62–3, 66
  crew's 30
  latent 6, 62–3, 66–7
  significant 14
fatal accidents 25, 57–8, 65–6, 126–7
  risk of 127–8
fatigue 18, 27, 30–31, 33, 138
field observations 26, 42–3, 113, 195
Flight Management Attitudes Questionnaire (FMAQ) 52
Flight Operations Quality Assurance (FOQA) 128, 137, 143, 196
Flight Safety Foundation 108, 132
focus groups 42, 45, 50, 52, 54, 85, 91, 113, 195, 197, 199–200, 205

Heinrich Ratio 57–8, 65–7, 82, 126–7
heparin 67–8, 75–6, 78, 121, 125
  dosage errors 8, 68, 75, 114, 119–21, 126
  see also Adverse Drug Event
heroes 12, 26, 43–5, 113, 132–3, 144, 147, 186, 188, 192, 215, 217–19
Herzberg 94, 101, 108
high-consequence industries 4, 11, 15–17, 43, 62, 99, 115, 148–9, 196, 205, 210, 226, 228, 231, 233
High Reliability Organizations (HROs) 17, 34, 98–100, 110, 142, 219, 225
hospital cost 160
Hospital Standardized Mortality Ratio (HSMR) 217

Hospital Survey on Patient Safety Culture (HSOPSC) 37, 61
human error management programs 110, 145, 190
Human Factors Classification System (HFACS) 25

improvements 19, 42, 60, 66, 88–9, 152, 160–61, 174, 185, 197, 203–4, 210–11, 220, 228
  systemic 17, 86, 88–9
incidents 1, 25, 27, 31, 57–8, 61, 65, 76, 88, 95, 107, 120, 163, 191, 221
  incident rate 217
infections 72–3, 77, 160, 210, 217
  centreline 76
  hospital-acquired 72–3, 135
information flow 92–4, 105, 198, 200, 204
innovation, culture of 211, 219
Institute for Healthcare Improvement (IHI) 20, 121, 124–5, 140, 174, 183, 185, 211–12, 217, 220
Institute of Medicine (IOM) 16–17, 20, 22, 124, 140
institutional identity 92–3, 105, 198
intermittent or isolated learning 5, 7–8, 113–14
International Civil Aviation Organization (ICAO) 81, 129, 137, 177, 179

Johns Hopkins Hospital 45, 60, 171–3
Joint Commission 18, 29, 68, 76, 79, 124–5
Joint Commission
  Accreditation of Healthcare Organizations 20, 124
  National Patient Safety Goals 120, 140
  Perspectives on Patient Safety 81
Joint Implementation Monitoring-Data Analysis Teams (JIMDATs) 127
Joint Safety Analysis Teams (JSATs) 127
Joint Safety Implementation Teams (JSITs) 127
Josie King Case 108, 171–5, 177
just culture 5–7, 17, 19, 23, 51, 58, 61, 76, 80, 93, 98, 108–9, 113, 179–80, 185

leaders, senior 162–3, 183, 212

leadership strategies 5, 12, 113, 115, 204, 215, 217–18
learning 7–12, 14, 20, 22–3, 39, 53, 59–61, 79–81, 105, 148–9, 151, 166–7, 193, 214–16, 228–33
  double-loop 14, 22, 176
  isolated 5, 7–8, 113–14
  organization's barriers 13
  single loop 14
  transformational 5, 7–8, 130, 173, 175, 191
  transformative 113–14, 175, 193
learning capabilities 13–14
  organizational 14, 22
learning organization 11, 22, 24, 38–9, 43, 53, 93, 110, 178, 190, 212
  continuous 8
learning organization questionnaire 23, 39
learning scale 5–8, 16, 21, 113, 179, 191
  continuous learning 5, 7–8, 14, 39, 113–14, 191, 193
  failure to learn 7
  intermittent or isolated learning 5, 7–8, 113–14
  transformational learning 5, 7–8, 130, 173, 175, 191
learning tools 11–12, 198
Line Oriented Safety Audit (LOSA) 26, 43–4, 128

M&M conferences 27, 53, 58–61, 67, 79–82
Maintenance Error Decision Aid (MEDA) 25
Maintenance Resource Management (MRM) 19, 27, 42, 44, 92, 96, 107, 110, 133, 138, 145, 181, 190
management commitment 39, 41
management, top 50, 101, 130, 180, 192, 200
medical errors
  preventable 16, 72, 126, 171
  preventing 33, 83
  self-perceived 70
medication error prevention 125–6
medication errors 20, 34, 67, 72, 75, 79, 82, 125 see also Adverse Drug Event (ADE)

mergers, corporate 168–9, 175
Methodist Hospital 75, 78, 117–18, 120–21
mission, organizational 38, 113, 117, 144, 184, 204, 215, 217–18
mistakes, honest 202–3
Model II 166–7, 173
  values 166, 173
Model III 166–7, 173–4
MRM programmes 129, 181–2

narrative analysis 147–50, 175, 178, 192
National Air Traffic Controllers Association (NATCA) 91, 138, 206
National Airspace System (NAS) 87, 89, 93, 102, 104, 195, 198, 202, 205
National Center for Patient Safety (NCPS) 20, 51
National-level safety performance goals 77
National Patient Safety Foundation 20, 109, 121, 124–5, 134–5
National Patient Safety Goals 125, 180
National Patient Safety and Quality Act 125–6
National Transportation Safety Board (NTSB) 27, 30–31, 33, 53, 64–7, 74, 82, 133, 137, 145, 184
non-punitive error 19, 180–81, 193, 195, 200–201, 205–6
non-punitive error-reporting system 105, 194, 205
  ASAP see also Aviation Safety Action Program
  ASRS see also Aviation Safety Reporting System
  PSRS see also Patient Safety Reporting System
North York General Hospital (NYGH) 184–5, 187, 210–12, 215–22, 225
Notice of Proposed Rule Making (NPRM) 129–30

organization
  Certified Patient Safety 126
  living 188
  reliable 225
  resilient 226–8
  safety-conscious 103
  ultrasafe 225

organizational
  accidents 23, 82, 109, 235
  change 19, 46, 176
  cultures 7, 23, 35, 53, 55, 74, 91, 139, 161, 166, 180, 193
  mission statements 115, 117
  performance goals 192
  resilience 225, 227
  transformation 46, 215
organizational factors 3, 27, 65, 92, 98, 101–2, 105–7, 198–9, 202
organizational learning 7–9, 14–15, 22, 24, 37, 46, 48, 52–3, 79, 176, 212–13, 227–9, 231, 233, 235
  continuous 180
  scale 114
organizational structures 2, 4, 49, 58, 71, 113, 117, 147, 153, 193–4, 229
  of hospitals 152–3
organizational values 4–5, 49, 115, 141, 171, 175, 178, 194, 232
  enacted 118
outcomes 1, 7, 9, 34, 58–9, 63, 71, 76, 88, 107, 140, 151, 155, 174, 189

PAC Model 106
pathways 28, 159–60, 189, 226
  intraoperative 157–8, 160–61, 177
patient care 22, 60, 69–70, 120, 150, 153, 162, 164, 172
patient safety
  culture 20, 37, 51, 54, 172, 174–5, 219
  goals 141
  heroes 134
  improved 17, 172, 210, 235
  organization value 162
  transform 174
Patient Safety and Medical Error Reduction 134
Patient Safety Organizations 79, 125
Patient Safety and Quality Improvement Act 78
Patient Safety Reporting System 7, 44
performance, organizational 48, 97, 141, 216
performance goals
  industry-level safety 58
  link unit-level safety 58
  measurable safety 192
  organizational safety 128
performance targets 77, 99, 127, 185
policies, organizational 45, 129
post-intervention safety climate studies 3, 25
procedures, published 4, 87, 116, 127
Professional Aviation Safety Specialists (PASS) 91, 138, 205
professionalism 36, 60, 86, 97–8, 105, 132, 138, 197–200, 203
  individual 19, 97–8, 109

Quality and Risk Management 217, 221
Quality and Safety in Healthcare 22, 24, 51, 53, 55, 80–81, 83, 234

reactive safety performance management strategy 76
relationships 11, 36, 38–40, 46, 53, 57, 63, 70–71, 87, 92, 94, 105–6, 152, 179, 198
  employee-management 86, 89, 97, 101, 181
  labour-management 91
  professional 94
reliability 3, 14, 17, 25, 27, 33–4, 37, 41, 54, 85–6, 90, 99–100, 102, 104–5, 107
reporting system 7, 19, 39, 74–5, 131, 142, 180–81, 190, 193, 195, 198, 200–201, 205–6, 231
resilience 100, 187, 226–8, 233
risk management 1, 20, 83, 100, 211, 217, 221, 227, 229
  proactive 229
Root Cause Analysis (RCA) 3, 25, 28–30, 58, 61–3, 79–83, 191, 231, 235
runway
  incursions 65–7, 76, 122, 128
  wrong 64–5, 74

safety
  current level of 88
  flight/patient 48
  nuclear 115–16
  systemic 19, 88
  value 150

Safety Attitudes Questionnaire (SAQ)
36–7, 41, 54, 140
Safety Climate 2–4, 12, 21, 24–6, 34–7,
39–42, 48, 50–51, 53, 85, 191,
193–5, 199, 203, 212
  measure 50, 85
  measurement 85, 191
  surveys 3, 35–6, 40–41, 50, 85, 107,
  131, 140, 191, 194, 199
safety culture
  assessment methods 25, 27, 29, 31,
  33, 35, 37, 39, 41, 43, 45, 47, 49,
  51, 53
  blame 181, 229
  changing 205
  compliance-based 19
  comprehensive understanding of 25,
  49, 205
  continuum 16, 21, 200, 210
  definition 5
  dominant states 5–6, 8, 200
  effective 110
  existing 2, 45, 49–50, 200
  improvement of 29, 37, 50, 95, 131,
  190, 200, 206
  maturity 185, 226, 228
  organizational 3, 27, 43, 71, 183
  robust 19
  scales 16
  state of 5–7, 16, 25, 50, 179, 195
  term 2, 85, 87
  transformation efforts 181, 188, 190
  transforming 23, 48, 109, 116, 179–85,
  187–9, 194–6, 234
Safety Culture in Aviation 19
Safety Culture in Healthcare 20
Safety Culture Pyramid Model 1–3, 5–7, 9,
11–13, 15, 17, 21, 25–6, 33–5, 41,
48–9, 57, 191, 212, 217–18
  see also Safety Performance, Safety
  Climate, Safety Strategies, and
  Safety Values
safety ethics 109, 138, 178
safety goals 39, 97, 113–14, 116, 138, 144
  organizational 194
safety index 34
safety leadership 4, 12, 24, 115, 144

Safety Management System (SMS) 4,
41–2, 77, 80, 129–30, 142, 179,
183, 191, 193
  programmes 130, 132, 142
  Safety Assurance 130–31
  Safety Policy 34, 44, 130
  Safety Promotion 130–31
  Safety Risk Management (SRM) 58,
  130, 142–3
safety measures 136
safety nets 63, 127
Safety Performance 2–3, 5–7, 24–6, 39–41,
50, 57–9, 61–3, 69, 77–9, 113–14,
129–30, 179, 191, 193–5, 210
  continuous improvement of 35, 130
  goals 76, 127–8, 132, 192
  improving 35, 48
  management strategy 58, 76
  targets 75, 77, 119
safety practices 4, 38, 44, 99
Safety and Quality Improvement Statement
45
safety risk assessment 77, 143
Safety Strategies 2, 4–5, 12, 21, 26, 34–5,
39–41, 50, 58, 113–15, 117, 139,
143–5, 191, 193–4
safety strategies maturity index 113–14
safety as a system property 140
Safety Values 4, 12, 21, 34, 40–42, 50, 147,
149, 151, 153, 161–3, 165–6, 173,
175, 191–3
  enacted 93
  espoused 93
  linking 50
safety values survey 39
secretive culture 5–6, 58, 113, 191, 198, 200
sensemaking of patient safety risks 79
Speed-Accuracy Trade-Off (SATO) 99, 155
stock values 73
  lost 73
stories 12, 14, 26, 43–5, 72, 117, 149–50,
171, 174–5, 177–8, 187, 218–19,
221–2
strategies 5, 7, 9–10, 20, 26, 38–9, 47–50,
62, 113–14, 127, 166–8, 173–5,
179, 191–3, 226–8
  assimilation 170

value-based 113
survey instruments 3, 25, 34, 36, 38–9, 42, 50, 140
    safety culture/climate 85
systems thinking 12, 15–16, 61, 212, 232, 234

team factors 92, 97, 101–2, 105–7, 198–9
Tech Ops Aviation Safety Action Program 201
thinking
    strategic 166–7, 174, 232–3
    value-focused 232, 234
Threat and Error Recognition Model 43, 58, 75
transformation
    value connection 48, 147, 171, 182, 187
    value conversion 48, 147, 171, 182–3, 186
    value creation 48, 147, 171, 175, 182, 186
transformational learning
trust 1, 6, 45, 47, 50, 75, 93, 96, 98, 161, 172–3, 187, 192, 197, 205
    interpersonal 74, 91, 93, 97–8, 148

underlying values 2, 12–13, 15, 45, 50, 62, 148, 153, 166–7, 173, 175, 185–6, 205, 224–6
undesirable events 6–7, 27–8, 31, 57–8, 76, 78, 115, 193
unions 91, 127, 138, 205
United Airlines 133, 145

unquestioned assumptions 2, 12–13, 45–6, 49–50, 115, 147–8, 167, 192

value congruence 192
value connection 48, 147, 171, 182, 187
value conversion 48, 147, 171, 182–3, 186
value creation 48, 147, 171, 175, 182, 186
values 4–5, 48–50, 92–3, 117–19, 147–50, 153–4, 157, 160–62, 165–6, 168–9, 173–5, 179, 182–4, 186–7, 191–2
    adversarial/defensive 173
    changing 53, 178, 190
    collaborative 173
    cultural 44, 170
    desired 187–8, 192
    employee 170–71
    enacted 5, 26, 48–9, 147, 149–50, 157, 165, 175, 188, 192, 194
    espoused 5, 26, 147, 150, 157, 165, 175, 186–7, 192
    group 131
    models 166, 175
    new 183, 188
    organization's 4, 115, 165
    personal 5, 40, 175
    professional 157
    shared 142, 147, 161, 187
    stakeholder 103–6
values-in-action 187
Ventilator Acquired Pneumonia (VAPs) 73
Vertical Patient Safety Perception Audit 162
Veterans Administration Patient Safety Culture Questionnaire 35

workarounds 4, 87, 99, 103, 198, 204